Register-based Statistics

Register-based Statistics

Registers and the National Statistical System

Third Edition

Anders Wallgren and Britt Wallgren
*Formerly at the Department of Research and Development at
Statistics Sweden*

Registered Offices
John Wiley & Sons, Inc., 111 River Street, Hoboken, NJ 07030, USA
John Wiley & Sons Ltd, The Atrium, Southern Gate, Chichester, West Sussex, PO19 8SQ, UK

Editorial Office
9600 Garsington Road, Oxford, OX4 2DQ, UK

For details of our global editorial offices, customer services, and more information about Wiley products visit us at www.wiley.com.

Wiley also publishes its books in a variety of electronic formats and by print-on-demand. Some content that appears in standard print versions of this book may not be available in other formats.

A catalogue record for this book is available from the Library of Congress

Hardback ISBN: 9781119632375; ePub ISBN: 9781119632689; ePDF ISBN: 9781119632641; Obook ISBN: 9781119632672.

Cover images: Courtesy of Anders Wallgren and Britt Wallgren
Cover design by Wiley

Set in 11/12.5pt TimesNewRomanMTStd by Integra Software Services Pvt. Ltd, Pondicherry, India
Printed and bound by CPI Group (UK) Ltd, Croydon, CR0 4YY

C9781119632375_160222

Contents

Index to Charts   xi

Preface   xix

Chapter 1   **Censuses, Sample Surveys and Register Surveys**   1
       1.1   The national statistical system   2
       1.2   The traditional census-based system   3
       1.3   New sources: Administrative registers and Big data   5
       1.4   Basic concepts and terms   7
           1.4.1   What is a register?   7
           1.4.2   Databases, records and observations   8
           1.4.3   What is a register survey?   10
           1.4.4   A register survey: The Income and Taxation Register 12
       1.5   New demands and opportunities require new methods   14
           1.5.1   A new paradigm is necessary   14
           1.5.2   New statistical methods   15
           1.5.3   The basic principles of register-based statistics 18
       1.6   Preconditions for register-based statistics   19
           1.6.1   Reliable administrative systems   20
           1.6.2   Legal base and public approval   21
           1.6.3   Political support to strengthen the statistical system 23

Chapter 2   **The Transition to a Register-based Production System**   25
       2.1   First obstacle: How to gain access to microdata?   26
       2.2   Protection of privacy and confidentiality   26
       2.3   Second obstacle: How to take care of dirty data?   29
       2.4   The new production process   30
           2.4.1   Contacts with administrative authorities   31
           2.4.2   Metadata have a new role   31
           2.4.3   Anonymisation of identity numbers   32
           2.4.4   Editing of a single administrative register   33
           2.4.5   Organising the work with administrative registers 36
       2.5   Third obstacle: The national registration system   37
           2.5.1   Legislation governs access to data   38
           2.5.2   Too many registers, but no good registers – what to do? 38
           2.5.3   Legislation rules obligations to report and what to report 40
       2.6   Why has the census been so important?   41

2.7   Creating the register system   42
    2.7.1   *Where do you live?*   43
    2.7.2   *Where do you work?*   45
    2.7.3   *With whom do you live?*   46
    2.7.4   *A centralised or decentralised national system?*   47
2.8   Register surveys and estimation methods   48
2.9   A traditional census or a register-based census?   49

Chapter 3   **The Nature of Administrative Data**   **51**
3.1   Comparing questionnaire and register data   51
    3.1.1   *A questionnaire to persons compared with register data*   51
    3.1.2   *An enterprise questionnaire compared with register data*   54
3.2   Enterprise registers for combined use   56
    3.2.1   *Corrections in accounting data*   57
    3.2.2   *Missing values in accounting data*   58
    3.2.3   *Administrative and statistical information systems*   59
3.3   Measurement errors in questionnaire and register data   60
    3.3.1   *Measurement errors*   61
    3.3.2   *Taxation errors*   62

Chapter 4   **Building the System – Record Linkage**   **65**
4.1   Record linkage   65
4.2   Record linkage in the Nordic countries   66
4.3   Deterministic record linkage   68
4.4   Creating variables by adjoining and aggregation   70
4.5   Probabilistic record linkage   73
4.6   Four causes of matching errors   79
4.7   The statistical system and record linkage   82

Chapter 5   **Building the System – Quality Assessment**   **85**
5.1   Four quality concepts   85
5.2   Making an inventory of potential sources   87
5.3   How can a source be used?   87
5.4   Quality assessment in a register-based production system   90
    5.4.1   *Analysing metadata*   91
    5.4.2   *Analysis and data editing of the source*   92
    5.4.3   *Comparing a source with the base register*   92
    5.4.4   *Comparing a source with surveys with*
          similar variables   93
5.5   Output data quality and quality of estimates   94
    5.5.1   *Analysing quality with a test census*   94
    5.5.2   *Analysing quality with samples from the new register*   95
    5.5.3   *Analysing quality with area samples*   96
    5.5.4   *Measuring quality of basic register variables with the LFS*   98
5.6   A coordinated system of registers   98
    5.6.1   *Are the base registers a coordinated system?*   98
    5.6.2   *Quality indicators at the system level*   99
5.7   Using the quality indicators   101

Chapter 6   **Building the System – Editing Register Data**   **107**
6.1   Editing in register surveys   108
6.2   Editing of a single administrative register   109
6.3   Consistency editing   110
    6.3.1   *Consistency editing – is the population correct?*   111
    6.3.2   *Consistency editing – are the units correct?*   118
    6.3.3   *Consistency editing – are the variables correct?*   120
6.4   Case studies – editing register data   121
    6.4.1   *Editing work within the Income and Taxation Register*   121
    6.4.2   *Editing work within the Income Statement Register*   123
    6.4.3   *What more can be learned from these examples?*   124
6.5   Editing, quality assessment and survey design   125
    6.5.1   *Survey design in a register-based production system*   125
    6.5.2   *Survey design – management problems*   127
    6.5.3   *Total survey error in a register-based system*   128

Chapter 7   **Building the System – The Population Register**   **129**
7.1   Inventory of sources   131
    7.1.1   *Time references*   131
    7.1.2   *Activities or 'signs of life'*   131
7.2   The Population Register based on full information   133
    7.2.1   *Object types – Changing and unchanging registers*   133
    7.2.2   *Variables with different functions in the system*   134
    7.2.3   *Updating the Population Register*   136
    7.2.4   *Registers and time*   137
    7.2.5   *Variables and time*   140
7.3   The Population Register in new register countries   140
    7.3.1   *Different systems of identity numbers*   141
    7.3.2   *Problems in countries without a central Population*
          Register   142
    7.3.3   *How to improve coverage of the Population Register*   143
    7.3.4   *Inventory of sources – addresses and time references*   146
7.4   Methods to measure and improve quality   148
    7.4.1   *Three kinds of surveys should be combined*   148
    7.4.2   *A new register-based system for statistics on persons*   150
7.5   Conclusions   151
7.6   Challenges in old register countries   152

Chapter 8   **The Population Register – Estimation Methods**   **155**
8.1   Estimation in sample surveys and register surveys   156
    8.1.1   *Estimation methods for register surveys that use*
          weights   157
    8.1.2   *Calibration of weights in register surveys*   157
8.2   Calibration of weights – the Swedish LFS   161
    8.2.1   *Use of auxiliary information in the LFS*   161
    8.2.2   *Nonresponse bias in the LFS*   162
8.3   Calibration – where do people live?   163
8.4   Methods to handle overcoverage   167

Chapter 9   Defining Register Populations – Coverage Errors   171
9.1   Defining a register's object set   172
    9.1.1   Defining a population   172
    9.1.2   Can you alter data from the National Tax Agency?   176
    9.1.3   Defining a population – the Farm Register   176
    9.1.4   Defining a population – integrated registers   178
9.2   Defining a calendar year population   179
    9.2.1   Defining a population – frame or register population?   180
    9.2.2   Sampling paradigm versus register paradigm   184

Chapter 10   Building the System – The Business Register   185
10.1   The Business Register and the National Accounts   185
10.2   The base register for economic statistics   187
10.3   The scope of the register and choice of object types   188
    10.3.1 The register with legal units and local units   189
    10.3.2 The register with enterprise units and kind of
             activity units   191
10.4   Inventory of sources   195
10.5   Creating and maintaining the Business Register   198

Chapter 11   The Business Register – Estimation Methods   201
11.1   Multi-valued variables   202
11.2   Estimation methods   205
    11.2.1 Occupation in the Activity and Occupation Registers   206
    11.2.2 Industrial classification in the Business Register   210
    11.2.3 Estimates from different register versions   213
11.3   Application of the method   214
    11.3.1 Change of industry and time series quality   215
    11.3.2 Transformation of weights   217
11.4   A decentralised or centralised statistical system?   218
    11.4.1 The Calendar Year Register and the National Accounts   219
    11.4.2 Choosing the best source for the National Accounts   220
11.5   Conclusions   224

Chapter 12   Censuses, Sample Surveys and Register Surveys –
Conclusions   227
12.1   Attitudes towards the register-based census   227
12.2   The new national statistical system   231
    12.2.1 The system of base registers   232
    12.2.2 Activity registers and longitudinal registers   234
12.3   Survey design   237
    12.3.1 Sample survey design   237
    12.3.2 Register survey design   238
    12.3.3 Creating register variables   241
12.4   Survey quality   245
    12.4.1 Quality of registers and register surveys   246
    12.4.2 The integration process – integration errors   247
    12.4.3 Frame errors   247

12.5	Organising the new production system	248
	12.5.1 Enterprise architecture and the register system	248
	12.5.2 The register system and data warehousing	249
	12.5.3 Missing values – a system-based approach	252
12.6	Final remarks	254
	12.6.1 The Statistical Population Register	254
	12.6.2 The system of base registers	255
References		**257**
Index		**261**

Index to Charts

Chapter 1	**Censuses, Sample Surveys and Register Surveys**	**1**
Chart 1.1	The Statistical System in a country can consist of the following actors	2
Chart 1.2	The Statistical System in a country can consist of the following surveys	3
Chart 1.3	Employment with census data	4
Chart 1.4	Employment with register data	4
Chart 1.5	The year of establishing new statistical registers in the Nordic countries	5
Chart 1.6	Example of a register and data matrix	9
Chart 1.7	A conceptual database model with three database tables	9
Chart 1.8	A database on individuals with three database tables	10
Chart 1.9	Two data matrices for different statistical purposes	10
Chart 1.10	From administrative registers to statistical registers – overview	11
Chart 1.11	From administrative registers to statistical registers – register processing	11
Chart 1.12	Different data sources for the Income and Taxation Register (I &T)	13
Chart 1.13	Employees by economic activity, November 2004, thousands	17
Chart 1.14	Register-based census statistics for one small municipality in Sweden	18
Chart 1.15	Four principles for using administrative registers for statistics	19
Chart 1.16	Two preconditions for using administrative registers for statistics	20

Chapter 2   **The Transition to a Register-based Production System**   **25**

Chart 2.1   The transition entails a fundamental change of statistical
               methods   25

Chart 2.2   An administrative register – unprocessed data in the
               input database   27

Chart 2.3   Corresponding statistical register – processed data in
               throughput database   28

Chart 2.4   Transforming administrative registers into statistical
               registers   30

Chart 2.5   Metadata system for input data from administrative
               registers   32

Chart 2.6   Identity database   33

Chart 2.7   Errors in profit and loss statements from limited
               companies, SEK million   35

Chart 2.8   Profit and loss statements for limited companies,
               SEK million   35

Chart 2.9   Registration of persons & demographic events in a
               Latin American country   38

Chart 2.10   Uncoordinated work with registers   39

Chart 2.11   Cooperation regarding the central population register   40

Chart 2.12   Sweden's system 1967–1984   43

Chart 2.13   Sweden's system 1985–2010   45

Chart 2.14   Sweden's system 2011   46

Chart 2.15   Survey design: reduce costs and/or improve quality   49

Chapter 3   **The Nature of Administrative Data**   **51**

Chart 3.1   Yearly turnover for the same enterprises in three sources,
               USD million   54

Chart 3.2   Distance for each ordered observation   55

Chart 3.3   Complete groups of enterprises   55

Chart 3.4   Three registers, Invoice Register, Client Register,
               Item Register   56

Chart 3.5   Statistical Sales Register for January 2012 – four
               transactions   57

Chart 3.6   Administrative Invoice Register   58

Chart 3.7   Statistical Invoice Register   58

Chart 3.8   Administrative Item Register   58

Chart 3.9   Statistical Item Register   58

Chart 3.10   Measurement errors – comparison of data collection
               methods   62

Chapter 4	**Building the System – Record Linkage**	**65**
Chart 4.1a	Matching without errors in the matching key (PIN)	68
Chart 4.1b	Matching with errors in the matching key (PIN)	68
Chart 4.1c	Matching with errors in the matching key (First name and Surname)	69
Chart 4.1d	Matching without errors in the matching key (PIN)	69
Chart 4.2a	The relations between persons, activities and establishments	71
Chart 4.2b	Wage sums for persons and establishments created by aggregation	72
Chart 4.2c	Industry and sex as derived variables for jobs created by adjoining	72
Chart 4.2d	Industry and number of employees as derived variables	73
Chart 4.3	Spelling errors, spelling variations and unstandardized addresses	74
Chart 4.4	Two ways of searching for duplicates in a Birth Register	76
Chart 4.5	Calculation of probabilities for outcomes for 1st name, 1st and 2nd surname	77
Chart 4.6	The eight outcomes for the matching key: 1st name, 1st and 2nd surname	78
Chart 4.7	Some outcomes for the matching key: Name and date of birth of the mother	79
Chart 4.8	Yearly turnover for the same legal units in three sources, USD million	80
Chart 4.9	Comparing gross yearly pay in quarterly and annual registers	81
Chapter 5	**Building the System – Quality Assessment**	**85**
Chart 5.1	A register survey: From administrative registers to statistical estimates	85
Chart 5.2	Statistical registers that are suitable for early development	87
Chart 5.3	Input and output data and the production process	89
Chart 5.4	The work with quality assessment of an administrative source	90
Chart 5.5	Indicators A1-A9 of input data quality – Relevance	91
Chart 5.6	Indicators B1-B7 of input data quality – Accuracy	92
Chart 5.7	Indicators C1-C5 of accuracy when comparing with the base register	93
Chart 5.8	Indicators D1-D4 of accuracy when comparing with related surveys	94
Chart 5.9	Coverage errors in the population register for Galapagos 2015	95

Chart 5.10	Coverage errors and area sample estimates of the population by district	97
Chart 5.11	Creating registers by linking different kinds of statistical units	99
Chart 5.12	Indicators E1-E4 of the quality of base registers in the system	100
Chart 5.13	Information from the administrative authority – relevance	101
Chart 5.14	Information from analysis and data editing of the source – accuracy	102
Chart 5.15	Information from integrating the source with base registers – accuracy	103
Chart 5.16	Information from integrating the source with related surveys – accuracy	104
Chapter 6	**Building the System – Editing Register Data**	**107**
Chart 6.1	Editing in sample surveys and register surveys	108
Chart 6.2	Automatic editing and imputation. Tax returns, 464 567 small enterprises	110
Chart 6.3a	The Business Register and the Annual Pay Register	112
Chart 6.3b	After matching the Business Register and the Annual Pay Register	113
Chart 6.3c	The statistical Annual Pay Register	114
Chart 6.4a	Combining the Business, Annual and Quarterly Pay Registers	115
Chart 6.4b	Final version of the statistical Annual Pay Register	116
Chart 6.5	All relevant sources should be considered simultaneously	117
Chart 6.6	Comparing gross annual pay in a quarterly and yearly source, microdata	119
Chart 6.7	The unit problem in business data	120
Chart 6.8	Comparing gross annual pay in QGP and AGP, microdata	120
Chart 6.9	Comparing gross annual pay in QGP and AGP, macrodata	121
Chart 6.10	The system of registers and surveys that was analysed	127
Chapter 7	**Building the System – The Population Register**	**129**
Chart 7.1	Decentralised but coordinated process to create registers on persons	130
Chart 7.2	The system of registers that constitutes the Population Register at the NSO	134
Chart 7.3	Different types of variables in the Population Register	135
Chart 7.4a	The Population Register at December 31, 2012	136

Chart 7.4b	Notifications regarding demographic events delivered 1 February 2013	136
Chart 7.4c	Old register matched with the new notifications	137
Chart 7.4d	Updated register 1 February 2013	137
Chart 7.4e	Three versions	137
Chart 7.5	Calendar year register for 2012	139
Chart 7.6	Events register for 2012 regarding change of address	139
Chart 7.7	Historical register regarding change of address	139
Chart 7.8	Longitudinal register for 2010–2012	139
Chart 7.9	Many registers, but no register that covers the entire population	142
Chart 7.10	The first steps towards a new statistical system	147
Chart 7.11	Different parts of the target population and the desired variables	148
Chart 7.12	Estimation of undercoverage	149
Chart 7.13	Coverage errors and area sample estimates of the population by district	150
Chart 7.14	Coverage errors and register-based sample estimates of population	151
Chart 7.15	A register survey: From administrative registers to statistical estimates	151
Chapter 8	**The Population Register – Estimation Methods**	**155**
Chart 8.1	Register of all persons in two small districts	157
Chart 8.2a	Employed persons by education and industry	158
Chart 8.2b	Employed persons by education and industry, missing values are shown	158
Chart 8.3	Persons by Education and Industry, adjusted for missing values	161
Chart 8.4	Nonresponse rates in the Swedish LFS	161
Chart 8.5	Nonresponse bias in the LFS, December 2011–2015. Relative bias, percent	162
Chart 8.6	Population estimates, Exercise 3	165
Chart 8.7	Exercises 1-5: Reducing coverage errors in population estimates	166
Chart 8.8	Estimates for Exercise 1 ('RC') and Exercise 5	166
Chapter 9	**Defining Register Populations – Coverage Errors**	**171**
Chart 9.1	Methodological differences between social and economic statistics	171
Chart 9.2	Frame populations are created before register populations	173
Chart 9.3	Undercoverage in an administrative register	177

Chart 9.4	Object sets when matching two registers	178
Chart 9.5	Calendar year register for the population in a (small) municipality	179
Chart 9.6	Calendar year register for 2013 for enterprises in a particular (small) region	180
Chart 9.7	The November frame and the Calendar Year Register 2004	181
Chart 9.8	Over- and undercoverage as number of legal units in the November frame	182
Chart 9.9	Errors due to undercoverage in the November frame 2004, SEK million	182
Chart 9.10	Undercoverage in November frame 2004, non-financial sector by industry	183
Chart 9.11	Overcoverage in SBS based on November frame 2004, by industry	183
Chapter 10	**Building the System – The Business Register**	**185**
Chart 10.1	The system of economic statistics	185
Chart 10.2	Inconsistent sources with wage sums in the National Accounts	187
Chart 10.3	Legal units by institutional sector and industry	190
Chart 10.4	Industry improved	190
Chart 10.5	Many economic activities	190
Chart 10.6	Different types of units in the Business Register	191
Chart 10.7	Legal units and enterprise units	192
Chart 10.8	Yearly turnover for the same legal units in three sources, USD million	192
Chart 10.9	Comparing gross yearly pay in quarterly and annual registers	193
Chart 10.10a	Each administrative system has its own object set – the effect on microdata	193
Chart 10.10b	Each administrative system has its own object set – different combinations	194
Chart 10.11	Sources with information for the Business Register	197
Chapter 11	**The Business Register – Estimation Methods**	**201**
Chart 11.1	Employees by industry, November 2004, thousands	202
Chart 11.2	Industry and number of employees as derived variables	203
Chart 11.3a	Calendar year register for the population of persons during 2013	204
Chart 11.3b	Average population 2013	204
Chart 11.4	Number of employed and wage sums in different registers	204

Chart 11.5   Register 1 continued – persons and weights   205
Chart 11.6   Employed by industry   205
Chart 11.7   Job Register with occupational data for six persons 206
Chart 11.8   Traditional register on persons with occupational
                information   207
Chart 11.9   Employed persons by occupation,
                traditional alternative 1   207
Chart 11.10   Register of combination objects:
                person • occupation   208
Chart 11.11   Employed persons by occupation according to two
                alternatives   208
Chart 11.12   Register on jobs of persons with occupational data 210
Chart 11.13   Persons and full-time employed by occupation, three
                alternatives   210
Chart 11.14a   Business Register year 1: Data matrix for
                establishments (local units)   211
Chart 11.14b   Business Register year 2: Data matrix for local units   211
Chart 11.14c   Number of employees by industry,
                traditional estimates   211
Chart 11.15   Data matrix with combination objects:
                local unit • industry   212
Chart 11.16   Number of employees by industry,
                estimated with combination objects   212
Chart 11.17   Four different registers belonging to the
                Business Register   213
Chart 11.18   Number of employees by industry   214
Chart 11.19   Employees by industry November 2004,
                thousands   215
Chart 11.20   Estimation of turnover within one industry
                using two methods   216
Chart 11.21   Turnover in an industry, two estimates   217
Chart 11.22   Transformation of weights   218
Chart 11.23   Calendar year register for 2007, legal units by sector
                and industry   219
Chart 11.24   Coverage errors in the Business Register
                compared with the Farm Register   220
Chart 11.25   Statistics Sweden's present way of using
                business surveys   222
Chart 11.26   A new estimation method based on
                integrated microdata   223

**Chapter 12   Censuses, Sample Surveys and Register Surveys –
                Conclusions   227**
Chart 12.1   Monthly population data for two Swedish
                municipalities 2000–2020   231

Chart 12.2   A register-based statistical system   232

Chart 12.3   Transitions from 2003 to 2004, men 25–54 years.
              % of each category 2003   236

Chart 12.4   Per cent employed after completing
              education 1987–1992   237

Chart 12.5   The three parallel processes with a register survey   240

Chart 12.6   Classification errors in the old and new Employment
              Register 1993   244

Chart 12.7   Model errors in the Employment Register 1993   244

Chart 12.8   Quality of estimates for two domains in the
              Employment Register   245

Chart 12.9   A longitudinal income register, data for six persons   246

Chart 12.10   Different parts of the survey production process   249

Preface

Survey sampling and register surveys – what is the main difference?
What is the main difference between a book on survey sampling and a book on register-based statistics? Both are books on survey methodology, but there is one important difference that should be understood from the outset:

- Books on survey sampling discuss *one* sample survey. We have *one* population and collect data for *one* survey. The sampling books then discuss how this sample survey can be designed in different ways.
- Instead, books on register surveys must have a systems approach. When discussing one register survey, we must also understand the role other registers play in the system of registers used when the statistical register in question is created and evaluated.

During the 1960s, Svein Nordbotten (1967) at Statistics Norway developed ideas on statistical information systems and explained that administrative sources should be used for statistical purposes. The *statistical information system* concept is quite different from the *statistical survey* concept, and the former is suitable for organisations that regularly collect data and produce statistics. He subsequently introduced what he called *statistical file systems*, or what we now call register systems used for the production of official statistics.

The register system is not just a set of registers. The system also has methodological implications regarding how it should be designed, coordinated and used. This was made clear in Statistics Denmark's book (Danish version 1994, English translation 1995). It explains how a system of statistical registers should be designed to produce a register-based population and housing census. The Danish book was the starting point for our work with registers at Statistics Sweden. The book is discussed in Sections 1.5.2 and 12.1.

Official statistics or corporate statistics?
The National Statistical Office is usually the largest organisation that collects data and produces official statistics in a country. However, statistical information systems are also important in other kinds of organisations. *Business intelligence* is a term that is often used for statistical information systems within corporations.

Bo Sundgren from Statistics Sweden was visiting professor at Linköping University during 1984–1986. He started a research programme on statistical information systems. At that time, we taught statistics to students of Business Administration and at the statistics programme in Linköping. We started our own research project 'Corporate information systems – Statistical analysis with the enterprise's administrative data'. We had contacts in this project with several companies, and we tutored many students who worked with papers for their degree. We recommend that university statisticians try something similar – this can be a good way to start statistical research on how to work with administrative data. Section 3.2 contains a short description of this area. If we had stayed in Linköping, perhaps we would have written books on corporate statistics. Instead, we went to Statistics Sweden, resulting in this book that is devoted to the national statistical system and how registers should be used to improve this system.

How did we work with this book?
Teaching, teaching, and teaching – this has been an integral part of our work with this book. We started with study circles for the teams working with different registers at Statistics Sweden. Together, we analysed the registers and discussed methodological problems.

We have continued along these lines since leaving Statistics Sweden – combining teaching and discussions with colleagues. We have visited some European countries as well as several countries in Latin America and the Caribbean, where our work was supported by the Inter-American Development Bank.

This approach has been very stimulating and has given us a broad picture of just where the important problems are. We have also learned how the subject area should be described and explained to statisticians in countries that are starting work with registers.

We hope that *Register-based Statistics – Registers and the National Statistical System* and its proposals will stimulate the discussion of statistical registers and register systems and provide support to those working with register surveys at national statistical offices.

Örebro, Sweden   Anders Wallgren
                                                       Britt Wallgren
                                                       ba.statistik@telia.com

CHAPTER 1

Censuses, Sample Surveys and Register Surveys

National statistical offices use three kinds of survey methodology when producing official statistics based on microdata: methods for *censuses*, for *sample surveys* and for *register surveys*. This book deals with the third kind of methodology – methods for register surveys, where instead of collecting data through interviewers and questionnaires, *administrative registers* from different administrative systems are adapted and processed to create *statistical registers* that are used to produce the desired estimates.

We introduce several concepts and principles that should be used when discussing register surveys. These concepts and principles form the methodological bases for this kind of survey. There is a growing interest in this area. Many countries increasingly use administrative registers for statistical purposes, and there is a growing demand for an understanding of register survey methodology.

However, preconditions differ – in some countries the preconditions are good, while in other countries there can be obstacles that make it difficult to use data from some administrative systems. We discuss such obstacles and how the national statistical system can be improved to reduce the problems. We give special attention to countries desiring to take the first steps towards a register-based statistical system.

The statistical offices in the Nordic countries started using registers during the 1960s; and experiences from these countries are important in understanding how statistical systems in other countries could be improved.

Purpose of this book
Our purpose is to describe and explain the methods that should be used for register surveys. Conducting a register survey means that a new *statistical register* for a specific subject matter is created with existing sources. The statistical register is then used to produce estimates required for the survey. What methods should be used in creating such a statistical register? One or more administrative registers are used when a new statistical register is created, and the statistical register can differ from the administrative sources in many ways.

Register-based Statistics: Registers and the National Statistical System, Third Edition. Anders Wallgren and Britt Wallgren.
© 2022 John Wiley & Sons Ltd. Published 2022 by John Wiley & Sons Ltd.

A *system of statistical registers* consists of a number of registers that can be linked to each other. In the Nordic countries, the national statistical offices have developed systems of registers that are used in the production of statistics. When new statistical registers are created, this register system becomes an important source that can be used together with different administrative sources. Another purpose of the book is to explain how such register systems should be designed and used in the production of statistics.

When a national statistical office starts using more administrative sources, the *statistical production system* of that office gradually changes. From a system based on enumerators or interviewers, address lists and maps, the system will become increasingly register based. Sample surveys will be based on the Population Register or the Business Register – variables in sample surveys can come from administrative registers as well as from telephone interviews or questionnaires. In addition to the change in methods used for sample surveys, new kinds of register-based statistics can also be produced. A third purpose of the book is to explain how administrative registers can be used to change the statistical production system of a national statistical office to improve cost efficiency and statistical quality.

1.1   The national statistical system

Official statistics in a country is produced by the national statistical system. We use two different interpretations of this system:

– The *system of actors* that is responsible for the official statistics.

– The *system of surveys* (censuses, sample surveys and register surveys) that these actors carry out to generate the desired microdata and estimates.

The actors in the system should cooperate to avoid duplicate work and conflicting or inconsistent surveys. The development of new register surveys requires that the national statistical office gains access to administrative registers that have been created and maintained by ministries or other administrative authorities. Data sharing and cooperation will then become necessary.

Chart 1.1   The Statistical System in a country can consist of the following actors:

1. The National Statistical Office[1] (NSO)
2. The National Advisory Council on Statistics
3. The Interagency Coordinating Committee on Statistics
4. The statistical offices of the Ministries
5. Statistical bodies of Regional Governments
6. Statistical bodies of Municipalities
7. Statistical bodies of public authorities

These actors interact with the political level that decides on legislation and funding of the national statistical system.

---

[1]We will use the abbreviation NSO in all chapters.

The national statistical system can also be defined as the system of all censuses, sample surveys and register surveys that is carried out by the actors in Chart 1.1.

Chart 1.2 shows several register surveys together with two examples of sample surveys. The thin lines between the surveys indicate that microdata can be linked with identity numbers.

The four registers in the grey circles play an important role in this kind of system. They define populations of different statistical units and link these populations with each other. These four registers are called the *base registers* in the system.

Chart 1.2   The Statistical System in a country can consist of the following surveys:

Sample surveys of persons	
Labour Force Survey	
Employment Register	Social insurance system
Income & Taxation Register	Preliminary taxation system
Education Register	Compulsory school, pupils
Longitudinal Income Register	Upper secondary school, pupils
Longitudinal Welfare Register	Register of University students
Registers for medical research	

Population Register — PIN — Activity Register

Address code   BIN EsIN

Geographical database, GIS
Land cover, Land use
Buildings register
Dwelling register
Register Real Estate — Address — Business Register

Value Added Tax Register
Payroll Register
Yearly Income Tax returns
Foreign Trade Register
Farm Register
Sample surveys of enterprises
Investment Survey

PIN=Identity number of persons   BIN=Enterprise identity number   EsIN=Establishment id number

Survey design refers to the development of methods that should be used for a specific survey. As a rule, the term is used for the design of sample surveys; but in subsequent chapters we discuss it with reference to the design of register surveys.

Survey system design is a term introduced by Laitila, Wallgren and Wallgren (2012) and describes the simultaneous work of improving or redesigning a system of surveys. For example, when a statistical population register has been developed in a country, all area samples of households can be replaced by samples of persons drawn from frame populations created with the new population register. This means that the whole system of household sample surveys is redesigned.

1.2   The traditional census-based system

Countries with a mainly traditional statistical system use interviewers to do population and housing censuses every ten years. Supported by maps and address lists, the interviewers go out and knock on all doors in the country. Sample surveys are conducted as a complement to the census, where interviewers go out and knock on an area sample of doors. The census is used to create the sampling frames.

Note that the *geographical location* of a person determines if that person is included in a sample or not. During the census or the sample survey interview, *all variables* required in the census or sample survey are collected by the interviewers. That is why concluding data sharing and cooperation agreements is not necessary – the different actors in the statistical system can manage their surveys independently. The interviewers can ask for names, birthdate and birthplace, but identities are not used in the production of statistics.

Census costs and value of information

A population and housing census is a costly and difficult operation, especially for developing countries. If the census has been successful, we obtain detailed information for small geographical areas and small categories of the population. However, the census estimates will be outdated after a few years. Using Swedish data, we can compare the information from a traditional census-based system with the information we obtain from a register-based statistical system.

Assume that we conducted a census in Sweden in 2001 and 2011 and that the next census will be done in 2021. Charts 1.3 and 1.4 show the number of employed persons in a municipality according to available data during February 2019.

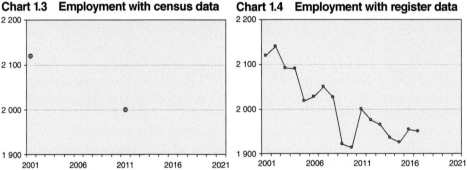

Chart 1.3   Employment with census data

Chart 1.4   Employment with register data

Chart 1.3 shows that there was a marked decline in employment between 2001 and 2011. But we do not know *when* the decline took place and we do not know what is happening *now*. We also know that the census involved substantial costs.

Chart 1.4 clearly shows the negative trend between 2001 and 2011. We can also see the effect of the financial crisis during 2009–2010. Finally, we can see what is happening now up to 2017. Chart 1.4 is based on estimates from Statistics Sweden's Employment Register and the costs are small compared with the costs for a census.

Remark: Administrative systems cover all time; censuses only give snapshots every tenth year.

In the early 1960s, the Nordic countries had statistical systems with area sampling and traditional population and housing censuses. Step by step, these countries managed to transition into completely register-based statistical systems; and in 2011 all countries conducted completely register-based censuses.

Chart 1.5 The year of establishing new statistical registers in the Nordic countries

Register	Denmark	Finland	Norway	Sweden
Population Register	1968	1969	1964	1967
Business Register	1975	1975	1965	1963
Dwellings Register	1977	1980	2001	2011
Education Register	1971	1970	1970	1985
Employment Register	1979	1987	1978	1985
Income Register	1970	1969	1967	1968
Population and housing census	1981	1990	2011	2011

Source: UN/ECE (2007)

We have visited many countries in Latin America and the Caribbean. All these countries had traditional census-based systems and the national statistical systems were decentralised. The national statistical office was responsible for the population and housing census, the agricultural census and some household sample surveys; the central bank was responsible for economic statistics and the National Accounts; and several ministries produced their own statistics.

All household sample surveys were based on area frames. Identities were not used when producing statistics and there was no tradition of data sharing and cooperation. If a country with this kind of statistical system wants to start using administrative registers for statistics production, then a comprehensive survey system redesign must be carried out.

1.3 New sources: Administrative registers and big data

Public authorities in all countries develop and maintain administrative systems that generate large volumes of microdata. When a national statistical office starts using administrative registers, a new production factor is added to the production system. The opportunities availed by the old methods of conducting a census or a sample survey remain, but new opportunities will be added to the production system when administrative registers are now used for statistical purposes.

The new administrative sources can often be difficult to use due to quality problems. However, it is still worthwhile to search for opportunities. In the future, new and better systems will be developed. National statistical offices should be active in the work to improve national administrative systems so that more and better systems can be used for official statistics.

Chart 1.5 clearly shows that the transition from the traditional census-based statistical system into the new register-based system took many years. All surveys changed when registers become available to produce official statistics.

– All censuses, sample surveys and register surveys will include identities. It will be possible to combine registers with other registers and to combine sample surveys with registers. Data can then be used much more efficiently, and this opens new possibilities for quality assurance and improvements.

– When different sources are compared, differences in coverage and differences between related variables are indicators of quality problems. The data sources need to be audited for quality.

– Censuses and sample surveys can use registers to generate frames. However, if the registers have coverage problems, it is wise to continue with area sampling. Area samples can then be used for quality assurance of registers.

Big data

Administrative registers are related to the more recent *big data* topic. Administrative data are considered by some authors as one kind of big data. However, we prefer to consider administrative data as a distinctive category, as they are much better structured and precisely defined in contrast to other types of data sources. In addition, administrative data has its own survey methodology, which is becoming established in more and more countries. If so-called 'big data' contain identities that can be linked to persons, areas, or enterprises, then big data are administrative data that can be linked with the system of statistical registers developed by the national statistical office.

Example: Toll payments and road sensor data

We sometimes drive over a new bridge and must pay a toll of 5 SEK ($\approx$ 0.5 USD) every time we use the bridge. Cameras read the registration number of our car, and we subsequently receive a mail based on the car owner's personal identity number with a request for a monthly payment for all the times we have used the bridge.

The request is based on a combination or system of three administrative registers: the register with camera data, the Vehicle Register and the Population Register. Thousands of registrations are made in this way every day and this is probably what many call 'Big data'. But the 5 SEK payment is an example of a tax payment and should be linked with the Income Register at Statistics Sweden, where the important variables *disposable income of persons* and *disposable income of households* are created for official statistics.

There are many examples of toll systems and road sensor data that collect data regarding activities connected with vehicles. Since the registration numbers of the vehicles can be linked to the owners' identities, all the data can be combined with other sources and used for statistical purposes.

All data in the register system in Chart 1.2 can be georeferenced and used to produce estimates for small geographical areas. These estimates can be supplemented with sensor and mobile phone data for the same areas.

1.4 Basic concepts and terms

The development of register-based statistics requires a common and rich register-statistical language. A common language within the theory of survey sampling is taken for granted. Terms such as frames, estimators and standard errors are well known and have a clearly defined meaning. Register-based statistics have the same need for well-established terms to stimulate the exchange of knowledge.

Two principles form the basis of this book – the *survey approach* to administrative data and the *system approach*. The survey approach involves the discussion of estimates, estimators and quality as in a book on sample surveys. The system approach builds on the *register system* concept. We also discuss the *production system* at a national statistical office and the role of administrative registers in the design and development of that system.

1.4.1 What is a register?

An administrative register is maintained to store observations on all objects to be administered; and the administrative process requires that all objects can be identified.

The following definition is valid for *administrative* registers:

An *administrative system* continuously generates *new data* to an administrative register; or it generates new administrative registers periodically.

An *administrative register* aims to include all the objects in a defined group of objects: the administrative object set. However, data on some objects can be missing due to quality deficiencies.

Data on the *object's identities* are used in the administration of objects. Therefore, the register can be updated and expanded with new variable values for each object.

Generation of new data, complete listing and known identities are therefore the characteristics of an administrative register.

Catalogue, directory, list, register, registry are different terms for the same concept. We use only the term *register*.

The following definition is valid for *statistical* registers:

A statistical register has been created by statisticians who use available administrative and statistical registers.

Complete listing and *identities* are also characteristics of statistical registers.

The administrative object set is replaced by the statistical concept *population*; and the known identities should be replaced by *anonymized identity numbers*.

The following are examples of registers:

- Civic, civil or national registration of the population in a country results in registers of citizens, births and deaths. This is an administrative system that continuously generates new data regarding the demographic events that affect the population.
- Income self-assessments from persons result in registers of all taxpayers for a given year. This is an example of an administrative system that generates new administrative registers yearly.
- In Sweden, enterprises with a turnover of SEK 40 million or more should report value-added tax monthly. This results in monthly VAT registers of reporting enterprises. For smaller enterprises, we obtain quarterly or yearly VAT registers. In all, we obtain three registers for three object sets: enterprises reporting monthly, quarterly and yearly.
- All export and import transactions are registered by Customs. Monthly registers are created with all transactions for a specific month. These transactions include identity numbers of exporting and importing enterprises.

The identities used in register processing can either be identity numbers that are unique within a national administrative system or an identity number in a subsystem with keys to the identities in other systems (as vehicles in the example with toll payments have links to the owners). It is also possible to use identities defined by, for instance, name, address, date of birth and place of birth.

1.4.2   Databases, records and observations

When Statistics Sweden migrated from mainframe computers to database servers, old terms such as *flat files* with *records* and *positions* were replaced by the term *database tables* with rows and columns.

We discuss these terms using the following example with data from an imaginary statistical register. Assume that we have a register containing data on all enterprises at a certain point in time. The number of objects in the register, illustrated in Chart 1.6, is given by N; and the register contains six variables.

In a *data matrix*, statistical data are sorted so that the matrix columns are the *variables*, and the matrix rows are the *observations* for the objects. The register in Chart 1.6 is represented by a data matrix with N rows and six columns.

Every statistical survey (census, sample survey or register survey) aims to create one or several data matrices containing *microdata*, which will then be processed for statistical purposes. The term *data matrix* can be considered a statistical concept for such a data set.

The columns in the matrix contain measurements of variables; the rows in the matrix contain *observations* for the objects in the register. The six-dimensional observation for Object 2 has been marked in grey in the chart.

Chart 1.6 Example of a register and data matrix

	Variable 1 Name of enterprise	Variable 2 Address	Variable 3 Business identity number	Variable 4 Turnover	Variable 5 Employees	Variable 6 Industry
Object 1	A's Painters	Address 1	BIN 1	12	9	F
Object 2	B's Bakery	Address 2	BIN 2	3	4	D
-	-	-	-	-	-	-
Object N	Z's Factory	Address N	BIN N	211	76	D

The old term *record* generates misunderstandings. It is a synonym to the term *observation* or *row* in a data matrix. But we often see that people use the term *administrative records*, and we suspect that they mean *administrative registers*. This becomes even more complicated in Spanish. The Spanish word *registro* stands for both *register* and *record*. We suggest that the term *record* should be replaced with the term *observation*. And if you mean *administrative registers*, state that and do not state *administrative records*.

The terms database and database table
Data matrices such as those shown in Chart 1.6 are stored in databases, which use various IT terms that are described here. In this book, the discussion is always on a conceptual level, namely how it is *thought* that registers and data matrices will *look*. Conceptual models often differ from the physical implementation of a database. Chart 1.6, for example, shows complete agreement between the register, the data matrix and the database table. However, this may not be the case in an actual systems solution.

The aim of a systems solution is that the database should be a flexible base for many uses in the register system. The structure of the data matrix can also be saved in a so-called *view* that shows you the data matrix that can differ from the way that the data has been stored.

Charts 1.7 and 1.8 illustrate different ways of describing the same database. Chart 1.7 shows the traditional way of describing a database. Chart 1.8 shows an example of a database's contents. The charts show the structure of the systems solution, i.e. how the data are physically stored. The database is *normalised* to ensure optimum consistency and space-effective storage.

Chart 1.7 A conceptual database model with three database tables

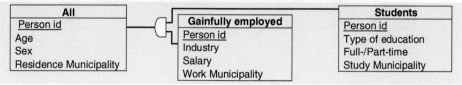

Chart 1.8   A database on individuals with three database tables

All

Person id	Age	Sex	ResMun.
PIN1	20	F	0586
PIN2	23	M	0586
PIN3	31	M	0586
PIN4	32	F	0586
PIN5	33	M	0586
PIN6	40	F	0586
PIN7	59	F	0586
PIN8	65	M	0586
PIN9	71	F	0586

Gainfully employed

Person id	Industry	Salary	WorkMun.
PIN2	G	52 000	0586
PIN3	G	287 000	0580
PIN4	A	193 000	0586
PIN6	D	291 000	0586
PIN7	D	314 000	0580

Students

Person id	Educ.Type	Full/Part-time	StudMun.
PIN1	AdultEduc	100	0586
PIN2	Univ	100	0580
PIN5	Univ	100	0580

Statistical processing for different aims can be carried out with the above database. Chart 1.9 shows an example of two different data matrices that could be created from this database. These data matrices can be created and stored physically, but they can also be formed temporarily during processing. A conceptual data matrix focuses on the specific problem to be solved, i.e. the statistical analysis to be carried out.

Chart 1.9 shows data matrices for employment and commuting in the way that statisticians require for their work. The data matrices in the chart would be called *denormalised* in database terminology, as several cells are missing values.

Chart 1.9   Two data matrices for different statistical purposes

Data matrix: Employment register

Person	Age	Sex	Employed	Industry	Salary
PIN1	20	F	No	null	0
PIN2	23	M	Yes	G	52 000
PIN3	31	M	Yes	G	287 000
PIN4	32	F	Yes	A	193 000
PIN5	33	M	No	null	0
PIN6	40	F	Yes	D	291 000
PIN7	59	F	Yes	D	314 000
PIN8	65	M	No	null	0
PIN9	71	F	No	null	0

Data matrix: Commuting register

Person	ResMun.	WorkMun.	StudMun.	Commuting
PIN1	0586	null	0586	0
PIN2	0586	0586	0580	1
PIN3	0586	0580	null	1
PIN4	0586	0586	null	0
PIN5	0586	null	0580	1
PIN6	0586	0586	null	0
PIN7	0586	0580	null	1
PIN8	0586	null	null	0
PIN9	0586	null	null	0

1.4.3   What is a register survey?

The original data are generated in public administrative systems. Definitions of object sets, objects and variables are adapted to administrative purposes.

Every authority carries out controls, corrections and other processing suited to their administrative aims.

When an authority delivers *microdata* to a national statistical office, further selections and processing may be carried out by the authority to meet the needs of the statistical office. The authorities also have *metadata* as definitions, administrative rules and information regarding quality, which are based on the administrative authority's experiences and investigations. This information is important for those receiving the data at the statistical office.

Producing statistics directly from the received administrative registers is generally not a good idea because these are not adapted to statistical requirements. The object sets, object definitions and variables must be edited. And as a rule, some processing will be necessary so that the register fulfils the statistical requirements for population, units and variables. The register-statistical processing, which aims to transform one or several administrative registers into one statistical register, should be based on generally accepted statistical methods.

Charts 1.10 and 1.11 show three important components of this work. We have found that people tend to use administrative concepts as they are. This can be acceptable in some cases – but in other cases it is unacceptable. The three issues of how to define population, units and variables of a statistical register are important for the quality of the statistics to be produced.

Chart 1.10 From administrative registers to statistical registers – overview

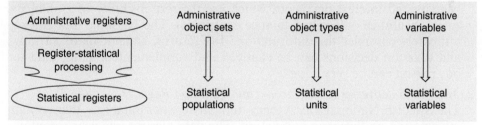

A statistical population or administrative object set consists of *N objects* or *units* or *elements*. Of these three synonyms, we as a rule use the term *object* for the units in an administrative object set and the term *statistical unit* for the units in a statistical population.

Chart 1.11 From administrative registers to statistical registers – register processing

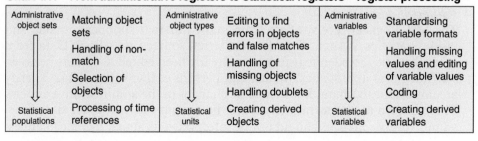

1.4.4   A register survey: The Income and Taxation Register

In this section we give an example of a register survey where Statistics Sweden created the population, statistical units and statistical variables so that the final statistical registers differ from the administrative sources used. About 40 administrative and statistical registers were used to create three statistical income registers. The register-statistical processing follows the steps in Chart 1.11.

How is the Income and Taxation Register created?

The Income and Taxation Register (I&T) is an important part of Statistics Sweden's register system. It describes income distribution and can be used for regional income statistics. It is also the basis for longitudinal income registers used by university researchers and a micro simulation model used by the Ministry of Finance.

This register utilises many administrative sources, and many administrative variables are used to create important statistical variables. In addition to these administrative sources, the register system at Statistics Sweden must also be used. The Population Register is used to define the population of the Income and Taxation Register, and important classification variables are imported from other registers in the register system to the Income and Taxation Register.

1. *Data generation at the National Tax Agency*
   The annual income self-assessment is based on tax returns from income earners and taxation decisions of the local tax authority. Both the income earner and the tax authority use statements of earnings for salary, sickness benefits and interest payments, which are the responsibility of employers, social insurance offices and finance companies. The National Tax Agency ultimately compiles this information. Tax returns, statements of earnings and taxation decisions can be changed and supplemented. Thus, data for one person can be very complex.

2. *Microdata deliveries to the Income and Taxation Register*
   The Swedish National Tax Agency annually creates databases that contain information on Sweden's population. The data files for one year – containing around nine million observations, each with around 300 variables – are delivered to the unit for income statistics at Statistics Sweden.

3. *Metadata to the Income and Taxation Register*
   Variable descriptions with names and definitions of variables accompany the deliveries from the National Tax Agency. Tax return forms, statements of earnings, taxation decisions and tax return instructions are also necessary for the correct interpretation of the data.

4. *Editing of data*
   The I&T Register receives data from many different suppliers outside and inside Statistics Sweden. Data from external sources are edited. Data from other Statistics Sweden registers have already been edited. Contacts between staff members at Statistics Sweden and persons at the National Tax Agency are important to obtain knowledge of changes in the administrative system. These in turn are important to ensure the quality of the register statistics – administrative changes should not be interpreted as actual income changes.

5. *Matching and selections*

Many registers should be processed to create the different sub-registers included in the Income and Taxation Register. Observations from different sources are matched using Personal Identification Numbers (PINs). Aggregation is carried out at the same time, i.e. all the statements of earnings data for a specific person are aggregated so that the person's income from work can be compiled. One type of processing is to select persons aged 16 and older who were also part of the population on 31 December.

6. *Derived objects are created*

A reference variable in Statistics Sweden's Population Register links each person to the *dwelling* where the person is registered as permanently living. All persons registered as permanently living in the same dwelling are classified as one *dwelling household*. These derived household units are created in the I&T Register and are used to produce income statistics for households.

7. *Derived variables are created*

Many derived income variables are formed. For instance, the wage or salary amounts are aggregated from the different earnings data to become a person's *income from work*. Every person's total income from work and capital plus transfer payments minus taxes becomes the person's *disposable income*. Variables are formed for households, such as *household type, number of consumption units* and *disposable income*.

Chart 1.12   Different data sources for the Income and Taxation Register (I &T)

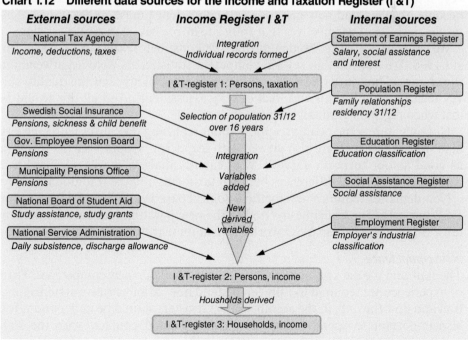

Chart 1.12 shows how the Income and Taxation Register receives administrative data from a variety of external sources and some Statistics Sweden

registers. The middle column shows the different phases when the sources are used in creating the new statistical register.

This example shows the importance of the principles in Chart 1.11. Statistics Sweden has access to many administrative registers with variables describing different kinds of income. The object set and the administrative variables have been processed to meet statistical needs. Many sources have been used to produce a statistical income register with rich content. The population in the income register is consistent with other statistical registers within the register system.

1.5   New demands and opportunities require new methods

The civic registers in the Nordic countries contain information on where persons and households have their permanent residence. This information is updated when persons change residence and household dwellings change. Thus, the administrative system generates daily data corresponding to the short form of the census.

This opens up new opportunities for timely population statistics and new kinds of social statistics. Longitudinal statistics based on these civic registers will also be something completely new for the national statistical office (NSO) – the fact that administrative systems cover all time provides new opportunities.

1.5.1   A new paradigm is necessary

The use of administrative systems instead of classical censuses and sample surveys represents a dramatic change for an NSO. Old competence and methods become obsolete and new methods and competence must be developed. This results in shifts in perspectives that translate into a new statistical paradigm.

We came to Statistics Sweden during 1989, just before the last classical census 1990. We noticed that several competent statisticians were negative about the change. The census was 'real statistical methodology' and the quality survey regarding the census was 'real quality assessment'. And all this would be lost when registers gradually replace the census!

Not only were some people sceptical about replacing the census with registers, data from tax forms were also viewed with scepticism. Could VAT tax forms be as good as our turnover survey? After careful analysis, it was admitted that there were only small differences between tax forms and the traditional survey.

New demands require a new methodological paradigm. Chart 1.15 in Section 1.5.3 presents four principles related to the new paradigm that are fundamentally important when an NSO works with administrative registers.

New competence

The staff and the national statistical office must build up new competence. Years of experience are required to fully understand how administrative systems function and how the data in the administrative registers should be understood. It is also important to spend time on establishing good contacts with the staff responsible for the administrative systems at the administrative authorities.

Managers must support this work of developing new competence and promoting discussions and seminars where the new sources and methods are discussed. To ensure that the administrative data are interpreted correctly, data should be analysed and compared with other sources. Staff working with register surveys should spend time on this kind of editing. Discussion of the analytical results is important so that the level of competence in interpreting administrative data improves for the whole staff.

1.5.2   New statistical methods

A paper by Hand (2018) was read before the Royal Statistical Society, where Hand discusses what should be included in a generally accepted theory for the analysis of administrative data. Quality, coverage, relevance, record linkage and protection of confidentiality are challenges mentioned by Hand, and these issues are discussed in the following chapters. Hand also points out that statistical textbooks often only mention random sampling and sampling errors. Since administrative data are not created by random sampling, students will obtain a limited view of statistics. Hand (2018) discusses the following challenges:

Data quality

Challenge 1. Statistics teaching should cover data quality issues.
Challenge 2. Develop detectors for particular quality issues.
Challenge 3. Construct quality metrics and quality scorecards for data sets.
Challenge 4. Audit data sources for quality.
Challenge 5. Be aware of time series discontinuities arising from changing definitions.
Challenge 6. Evaluate the impact of data quality on statistical conclusions.

'Data = all'?

Challenge 7. Explore potential sources of non-representativeness in the data.
Challenge 8. Develop and adopt tools for adjusting conclusions in the light of the data selection processes.

Answering the right question

Challenge 9. Explore how suitable the administrative data are for answering the questions. Identify their limitations and be wary of changes of definitions and data capture methods over time.

Challenge 10. Report changes and time series with appropriate measures of uncertainty, so that both the statistical and the substantive significance of changes can be evaluated. The measures of uncertainty should include all sources of uncertainty which can be identified.

Causality and intervention

Challenge 11. Be aware that administrative data are observational data, and exercise due caution about claiming causal links.

Combining data from different sources

Challenge 12. Be aware of the risks that are associated with linked data sets and the potential effect on the accuracy and validity of any conclusions. Recognise that quality issues of individual databases may propagate and amplify in linked data. Develop better measures of overall combined data quality.

Challenge 13. Continue to develop statistically principled and sound methods for record linkage and evidence assimilation, especially from non-structured data and data of different modes.

Challenge 14. Develop improved methods for data triangulation, combining different sources and types of data to yield improved estimates.

Confidentiality, privacy and anonymisation

Challenge 15. Continue to explore anonymisation and deidentification methods.

Regarding Hand's Challenges 12 and 13, our experience from Statistics Sweden tells us that there are no risks of linking data sets with data on persons if the identity numbers are of Nordic quality. The best method to develop sound methods for record linkage is to improve the system of national identity numbers – probabilistic record linkage is not the solution. We return to this topic in Chapter 4.

New and old paradigms

Sample surveys are based on methods derived from an established theory: *sampling theory*. This theory has been developed within the academic world and statistical offices and consists of generally well-known terms and principles. Scientific literature and journals develop and spread the methodologies for sampling and estimation. Because the terms and principles are well known, people working with sample surveys can easily communicate and exchange their experiences.

Censuses with their own data collection are based on a long tradition of population and housing censuses. Measurement errors, design of questionnaires, coverage errors and nonresponse are methodological issues that also apply to sample surveys. Censuses and sample surveys are closely related in terms of methodology – censuses are often considered as special cases where the sample is the entire population.

Although register-based statistics are a common form of statistics used for official statistics and business reports, no well-established theory in the field exists. There are no recognised terms or principles, which makes the development of register-based statistics and register-statistical methodology all the more difficult. As a consequence, *ad hoc methods are used instead of methods based on a generally accepted theory*.

One important reason for this shortfall is that the subject field of register surveys is not included in academic statistics. Statistical theory within statistical science is understood as consisting of probability theory and statistical inference. Sampling theory is included within this theoretical school of thought, but register surveys based on total enumeration are not.

Unfortunately, statistical science has so far not included any theory on statistical systems. Statistical offices, larger enterprises and organisations do not often carry out separate surveys. The building of *statistical information systems* is more common, and these constantly generate new data. A statistical theory is necessary to describe the general principles and to develop the conceptual apparatus for such statistical information systems. Register surveys should be included in this theory.

When the system approach is missing

Chart 1.13 illustrates a common problem. Estimates from four different surveys are compared, and these comparisons show clearly that the system approach is often missing in the work with statistical surveys. People are fully occupied with their own surveys and different surveys are published at different points in time. As a rule, most estimates are unique for one survey. However, in Chart 1.13 we found one identical variable and created the table with corresponding estimates from each survey.

If we look at each survey separately, we do not see any errors except for the sample survey in (4) where we have margins for the sampling error. But when we look at the four surveys together, we understand that there must be more serious errors in these surveys. We thus need *a theory for systems of surveys and new methods for quality assessment.*

Chart 1.13   Employees by economic activity, November 2004, thousands

Economic activity	Business Register Enterprises (1)	Establishments (2)	Employment Register (3)	Labour Force Survey (4)	Error margin (5)
Agriculture, forestry, fishing	35	37	37	26	5
Mining, quarrying, manufacturing	688	636	717	640	23
Electricity, gas, and water	21	22	28	29	5
Real estate, business activities	457	524	457	470	20
...	...	...	...	...	...
Education	382	408	431	462	20
Health and social work	836	684	675	675	24
Other service activities	142	163	175	168	13
Unknown activity	0	0	38	4	
Total	**3 763**	**3 763**	**3 924**	**3 778**	**43**

Why are there such large differences between the surveys? The estimates for mining, quarrying and manufacturing can be 636 000 or 717 000 – the inconsistencies are more serious than the sampling error.

The methodological work should have included the following three steps: (1) compare surveys and find errors and inconsistencies; (2) find out why we have these inconsistencies; and (3) reduce the errors and inconsistencies.

Chart 1.13 also illustrates that we only have one established way of giving a numerical description of the quality of published estimates – margins for the sampling error. There is no commonly used way of describing the quality of register-based statistics. However, the non-sampling errors of sample surveys are as a rule not described in the same clear manner as the sampling errors. We also lack methods here for giving a numerical description of the quality of published estimates.

The system approach – survey system design
In 1995, Statistics Denmark published *Statistics on Persons in Denmark – A Register-based Statistical System.* The Danish book presents a systematic review of register-statistical work and describes how to design a well-prepared register system. The book was the first attempt to create a theory for register-based statistics and describe the methods used.

Statistics Denmark developed a system of coordinated and consistent register surveys to enable replacement of the traditional census with a register-based census. Hence, an entire system of register surveys was developed instead of creating a number of independently developed surveys. Statistics Denmark used the traditional census as a model and created registers that could produce population statistics, household statistics, employment statistics, education statistics and income statistics equivalent to the corresponding census statistics.

This was probably the first example of survey system design. The pioneer work done by Statistics Denmark constitutes a milestone in survey methodology.

To understand the significance of Statistics Denmark's achievement, Chart 1.13, which illustrates the lack of a theoretical approach, should be compared with Chart 1.14 based on the new register-statistical theory. Several registers were developed to produce yearly census-like statistics for small areas and small categories.

Chart 1.14   Register-based census statistics for one small municipality in Sweden

Population Register		Employment Register		Education Register					Income Register			
Age	Number of persons	Em-ployed	Not em-ployed	Com-pulsory	Upper secon-dary	Post secon-dary	Post-graduate	Not known	Yearly earned, SEK thousands (1 USD ≈ 8 SEK)			
									0	1-139	140-279	280-
0–15	1 358	-	-	-	-	-	-	-	-	-	-	-
16–19	384	84	300	297	67	0	0	20	120	259	5	0
20–24	328	197	131	55	233	31	0	9	38	153	105	32
25–34	719	622	97	88	423	201	1	6	15	141	320	243
35–44	962	846	116	83	554	319	2	4	19	106	381	456
45–54	910	776	134	171	500	226	4	9	24	89	363	434
55–64	1 071	791	280	295	491	271	7	7	21	145	468	437
65+	1 402	138	1 264	758	438	194	8	4	1	508	719	174

1.5.3   The basic principles of register-based statistics

We have formulated four basic principles for using administrative registers listed in Chart 1.15. We use these principles in the book and gradually introduce the register-statistical terms necessary for the discussions.

In a traditional statistical system with censuses and sample surveys, data are collected for each survey and all variables needed for each survey are simultaneously collected. This means that the established tradition is: *one* survey with *one* data collection gives *one* set of data for *one* specific purpose. Population and variables have been determined for this specific purpose.

Chart 1.12 with the Income and Taxation Register shows that none of this needs to apply to a register survey. Many sources are combined to achieve good coverage and to include all kinds of income. With these sources, three statistical registers are created for different purposes. According to the *Transformation principle*, all relevant sources should be combined when a statistical register is created, and population and variables in this register may differ from the sources.

Chart 1.15 Four principles for using administrative registers for statistics

---

1. Transformation principle

Administrative registers should be transformed into statistical registers.
All relevant sources should be used and combined during this transformation.

2. System principle

All statistical registers should be included in a coordinated register system.
This system will ensure that all data can be integrated and used effectively.

3. Consistency principle

Consistency regarding populations and variables is necessary for the coherence of estimates from different register surveys.

4. Quality principle

The register system should be used for quality assessment of statistical surveys based on microdata comparisons with other surveys in the production system.

The *System principle* emphasises the importance of the register system as an important source, but registers in the system should follow the *Consistency principle* to avoid conflicting information from the different registers. The work with registers must be coordinated to achieve consistency. Populations and variables must be defined to suit the system, not individual surveys.

The *Quality principle* stresses that quality assessment must be done in a new way. In sample surveys, we describe the quality by computing confidence intervals. But how should we work with quality in register surveys? The quality issue is discussed in Chapters 5 and 6.

The goal: A coordinated statistical system

All microdata can be integrated in a national statistical system with a coordinated register system. In a coordinated system, the populations in the different registers are consistent regarding coverage, and the variables in different registers do not have contradicting values.

Of course, perfect consistency is difficult to obtain, but if coverage problems and variable inconsistencies are tolerable, we can say that the system is coordinated. Integration of microdata will become possible without disturbing problems regarding lack of consistency between populations and variables.

1.6 Preconditions for register-based statistics

Preconditions differ between countries for sample surveys, censuses and register surveys. Hence, the preconditions for statistical methods are different. The choice between cluster sampling and one-stage sampling depends on whether there is a Population Register or if you must use maps and address lists. Regression estimation and calibration are methods that depend on the number and quality of available register variables. This means that an *increased use of administrative registers will change the preconditions for all kinds of surveys.*

For register surveys, the differences between countries are even more significant. Legislation on national registration and the taxation of persons and enterprises determines the character of the administrative systems used in each country. The legislation regarding statistical production and protection of statistical data also differs. Therefore certain methodological issues are important in some countries but not in others.

The two main preconditions for using administrative registers for statistical purposes are stated in Chart 1.16.

Chart 1.16   Two preconditions for using administrative registers for statistics

> *5. Identity number principle*
>   Unified systems of identity numbers are used in all administrative systems. The same identity number should follow an object over its lifetime.
> *6. Legal principle*
>   A statistical office should have access to administrative registers kept by public authorities. This right should be supported by law as should the protection of privacy.

1.6.1   Reliable administrative systems

Reliable administrative systems will generate data of good administrative quality. Good administrative quality is a necessary but not sufficient condition for good statistical quality. The systems for tax administration and welfare programmes will gradually develop and change, and these changes will determine what administrative data can be used for future statistical purposes. Therefore, maintaining close and long-term relations with administrative authorities and politicians is important for national statistical offices.

The long-term strategy requires high-level contacts to promote strategic changes that will improve statistics production. The statistical office must explain to the administrative authorities how their data are used for statistical purposes. The statistical office also needs detailed information on how the administrative systems are organised and what changes are planned. Close and long-term contacts at all levels are required for these purposes.

What aspects of national administrative systems are important for statistical offices? We note two such aspects here, coverage and identity codes.

Coverage – the systems should cover all
The Nordic systems for child benefits are good examples of coverage. All children in defined age groups are entitled to a sum of money. All parents want the entitlement – but to receive the money, the parents must be registered as parents to the child in question and national identity numbers are required for the parents and the child. The residential address in the administrative register must also be correct so that the parents can receive the money. This system covers all children and all parents. All persons in the country will gradually be covered as the information in the system's registers is maintained and updated. The register will contain administrative, but also statistically important, links between all parents and children.

Good coverage requires that the administrative systems cover both urban and rural populations, rich and poor citizens, and small and big enterprises. The ideal is no selectivity. If suitable methods are not developed, selectivity will result in biased statistical estimates. For instance, in the Nordic countries all seriously ill persons will see a doctor, and all doctors know that cancer patients should be reported to the National Cancer Register. In this way we can be almost certain that all patients with a cancer diagnosis are in the Cancer Register. If rural or poor persons are underrepresented, estimated cancer incidence and mortality figures would have low quality.

Unified systems of identity codes

Identities are important in administrative systems. Legally important relations between persons, such as husband and wife, or parents and children, are registered with the identities of the persons in question. In many registers, the legally important relations between owners and different kinds of property are recorded with both the identities of the owners and the property. For taxpayers, it is important that the tax paid is recorded together with the identity of the taxpayer. Therefore, it is in the interest of each taxpayer to use a correct identity in each transaction. The legal importance of identities explains why identity data as a rule are of high quality in many administrative registers.

The best way to handle identities in administrative systems is to use *national identity numbers*. Persons, enterprises and property should be assigned unique identity numbers that are used in all administrative systems in the country. The same number should follow each person, enterprise or property over its lifetime.

Not only will administration become efficient; the statistical production system will become efficient. When administrative data are used for statistical purposes, it will be possible to link observations and create important statistical comparisons. Record linkage will be easy with unique national identity numbers, and the risk of false matches and false non-matches will be low. The statistical possibilities that national identity numbers create are explained in the following chapters.

It is advantageous if the identity numbers have no relation to any attributes of the objects that are to be identified. For example, identity numbers for persons should not depend on name, sex, or address of the persons, because such attributes can change over time. We use throughout the abbreviation PIN for national identity numbers for persons and BIN for national identity numbers for legal units representing enterprises.

1.6.2 Legal base and public approval

There are preconditions concerning legal base and public approval that enable the efficient use of administrative registers for statistics. These preconditions are discussed in UN/ECE (2007) and we build on that discussion here.

Chapter 2 describes the transition from a traditional statistical system into a register-based system. An important part of the work with this transition is to improve the preconditions for a register-based system. The Nordic countries have spent several years working hard to improve these preconditions.

Legislation determines what data are generated

The national administrative systems for taxation and welfare are based on legislation that determines the kind of administrative data generated within these systems. For example, if citizens pay income tax to municipalities, then the authorities must know where each citizen lives. The municipal taxation and welfare systems are the legal base for the Nordic administrative population registers. They are used not only for taxation and municipal welfare, but also for elections where the population register defines where each voter votes. For statistical purposes, this creates very good links between persons and geography that facilitate regional statistics. The administrative registers are updated every day, which makes possible timely monthly demographic statistics.

Legislation to improve the national statistical system

Politicians want to reduce the response burden of persons and enterprises as well as the direct costs for the production of official statistics.

- Legislation should provide the national statistical offices access to administrative microdata, including identities, and the right to use the data for official statistics and research.
- Legislation should provide statistical offices the authority to match data from different sources, and use data that were not originally generated for statistical purposes.
- Legislation could also instruct statistical offices to use data from administrative registers first, and to conduct sample surveys or censuses only if available administrative data are insufficient.
- Some laws have the sole purpose of making register-based housing and population censuses possible. For example, the Nordic parliaments have decided that all employers must provide information on where all employees work – the establishment address for all. This information is provided by income statements with data on employer identity, establishment identity, employee identity and wages, preliminary tax and social insurance paid. These income statements play an important role in the Nordic statistical systems, since we obtain important links between three different object types. The parliaments have also decided that all persons should be registered at the dwelling where they live. Then it will be possible to create statistics for households defined by the common dwelling in the register-based census.

Legislation on data protection

According to the second precondition in Chart 1.16, a national statistical office should have access to administrative registers kept by public authorities. This right, as well as the protection of privacy, should be supported by law. Improved legislation will take time – in the short run, agreements with ministries and public authorities will be necessary.

Legislation that gives a statistical office access to administrative data is discussed above, and the protection of privacy and confidentiality are discussed below.

The principle of *one-way traffic* is important for data protection. Microdata can go from administrative authorities to the statistical office but never in the reverse direction. The legislation on data protection should rest on a reasonable balance between protection of integrity on the one hand and increased costs and difficulties for statistics production on the other. An important task for top management at a national statistical office is to explain the consequences generated by proposed legislation to lawyers and politicians.

Public approval

The cooperation between register authorities and national statistical offices should be open and transparent. The fact that administrative data are used for statistical purposes should not be kept quiet; instead, the benefits of using administrative data and the efforts to protect integrity should be explained in open discussion and public debate.

It is important to explain that individual observations regarding persons are anonymous in statistics production, which is in contrast to how administrative authorities handle the same data.

If the national statistical office has a good reputation as trustworthy, gaining access to administrative data for statistics production will be easier. However, one mistake in the protection of integrity can immediately destroy this reputation.

Persons and enterprises do not want to be required to report to both an administrative authority and the national statistical office. Not having to do so will make public opinion more favourable to the use of administrative data for statistical purposes. The double provision of data becomes difficult to motivate – why respond to a questionnaire on the enterprise's turnover when you also submit a value-added tax return to the Tax Agency that includes the same information?

1.6.3 Political support to strengthen the statistical system

Political support and consensus is crucial for developing and strengthening the national statistical system. The national statistical office must have a role of coordinating and regulating the country's official statistics. This role must be recognised by ministries and other authorities that belong to the statistical system. The following quotation from Dargent et al. (2018) is a reminder that this is not always the case.

'It is therefore noteworthy that incentives for strengthening statistical capacity are not always aligned within government or, in other words, there are trade-offs affecting the development of statistics policies. For example, governments need data for better decision making, but these same data can be used as a tool for citizens to demand accountability from governments for their decisions and thereby limit their discretionality. Consequently, the best and most transparent statistics policy can sometimes run counter the government's interests, creating incentives to ensure that these offices remain weak.'

Source: From Dargent et al. (2018) Who wants to know? The Political Economy of Statistical Capacity in Latin America. Inter-American Development Bank.

CHAPTER 2

The Transition to a Register-based Production System

The transition from a census-based into a register-based national statistical system will change the entire production system. The new system of statistical registers will be used for social and economic statistics. Note that once a statistical office starts to use administrative registers for some issues – not only will the census be transformed – the entire statistical system will gradually change completely.

This chapter treats some of the main issues that become important when the national statistical office (NSO) in a country starts to systematically work with administrative registers. Subsequent chapters describe the register-based system in more detail as well as survey methods for register-based statistics. Chart 2.1 describes methodological differences between the two kinds of systems.

Chart 2.1 The transition entails a fundamental change of statistical methods

A census-based system	A register-based system
All surveys use area frames and maps or address lists.	All sample surveys use frames that have been created with registers.
The Population and Household Census is used as sampling frame for household samples and can be used as frame for the Agricultural Census.	The Population Register is used to create frames for samples of persons. The Farm Register and the Business Register are used to create frames for samples of holdings, enterprises or establishments.
Each survey stands alone, cooperation and data sharing are not necessary.	Cooperation and data sharing within the NSO and with ministries and authorities are essential.
Identities are not used.	Identity numbers are used in all registers and surveys. Identity numbers are very important in the production process. The identities should be anonymised.
All variables are collected at the same time during the interview or when a questionnaire is completed.	Variables come from different sources and from different times. A statistical register as a rule contains variables from several registers. A sample survey contains a combination of variables from registers and variables collected with a questionnaire.
New data must be collected by sending out questionnaires or interviewers.	New data are constantly or periodically being generated by administrative systems.

Register-based Statistics: Registers and the National Statistical System, Third Edition. Anders Wallgren and Britt Wallgren.
© 2022 John Wiley & Sons Ltd. Published 2022 by John Wiley & Sons Ltd.

Section 1.6 describes the preconditions for register-based statistics. This chapter describes how to improve the preconditions through long-term projects and cooperation. This is the required hard, long-term work to enable the NSO to use administrative registers efficiently. The Nordic countries have spent much work with this. We have visited several countries that have started this work, and we find similar problems in all countries. Progress will take time, and it is wise to have a long-term strategy to improve the statistical system as well as to develop methods to use the registers available today.

2.1   First obstacle: How to gain access to microdata?

The starting point for a national statistical office could be that the NSO is responsible for the Population and Housing Census, several area-based household sample surveys and some business sample surveys. The NSO has not gained access to administrative registers with identity numbers, but it has access to aggregated administrative data and some business data without the real identities.

Ministries and other national authorities have their own administrative registers with names, addresses and identity numbers. However, there is a long tradition of 'protecting confidential data' that explains why the NSO does not gain full access to the administrative registers. When there is lack of cooperation, duplication of work can occur and conflicting statistics can be published by different actors in the national statistical system.

When an NSO wants to start using administrative registers for statistics production, political support is required for the long-term work of improving legislation regarding the right to get access to administrative registers with identity numbers. In addition, there must be consensus among ministries that cooperation and data sharing are necessary for improving the national system, and that the NSO should have a leading role in this work with developing the statistical production system.

To get this support, it is important that the NSO explains to all other actors that it will use administrative registers in a way that protects confidentiality. An IT system should be developed, where real identity numbers are replaced by anonymous numbers, and the NSO should explain that only these anonymous numbers will be used in the production of statistics. Until the new legislation is completed, agreements could be made with ministries and other authorities with registers so that the NSO can start the work with improving the national statistical system. The fact that names and official identity numbers are not used in the production of register-based statistics will facilitate completion of these agreements.

2.2   Protection of privacy and confidentiality

It is important that statistical offices maintain high standards for protecting the privacy of people and enterprises. Even if the real risk of threatening people's privacy with statistical use of administrative data is comparatively small, a debate in today's media could have serious consequences regarding the reputation of a statistical office.

We use the term *confidentiality*, which applies to data regarding individuals and enterprises. When a register system is developed to facilitate the integration of sources with the aim of raising the quality of official statistics, the protection of confidentiality can be improved in the following ways:

- Official identification numbers are replaced by anonymous numbers. This can be handled by a small group at the NSO or by a special authority.[1]
- Existence of variables with text is minimised.
- Exact birth dates and residential addresses are only stored in the Population Register, with restricted access.
- The risk that data regarding persons or enterprises can be derived from statistical tables in publications or official databases should be minimised.
- Before researchers, after application, are granted the opportunity to analyse data matrices with microdata, these data matrices should be processed to minimise the risks of disclosure of information on individuals.

Data submitted to statistical offices to produce statistics are covered by statistical confidentiality in each country's legislation. In addition to this legislation, data processing within a statistical office should be organised so that confidentiality is efficiently protected.

No text and no official identities in databases for throughput and production

The administrative data delivered to a statistical office can contain names, addresses and other text data which the administrative authority needs for administrative purposes. Text information as addresses and names of well-known persons and enterprises draws attention to itself and must be removed when data has been delivered to the statistical office. These details should be replaced by codes when carrying out the preliminary register processing. This is illustrated in the fictitious example in Charts 2.2 and 2.3.

Chart 2.2   An administrative register – unprocessed data in the input database

PIN	Name	Address	Post Code	Enterprise, establishment	Job title	Occupation code, TNS	Actual salary	Extent of work
560230-1234	Pson Per	1st Street 7	111 11	Statistics Sweden, Stockholm	IT specialist	4321	18340	0.60
670631-2345	Ason Eva	2nd Street 2	777 77	Statistics Sweden, Örebro	Head of department	1234	45780	1.00

As part of the data editing, official personal identification numbers (*PIN*) should be replaced by anonymous identification numbers (*anPIN*); all plain language should be replaced by codes; and all variables that are not of statistical interest should be deleted. In addition, statistically important variables are imported from other registers and derived variables are formed. Also, the

---

[1] In Austria there is a special authority that creates anonymous identity numbers before the NSO gets the data.

identities of enterprises and establishments are replaced by anonymous numbers *(anBIN* and *anEsIN)*.

The names and identity numbers of the persons, enterprises and establishments in Chart 2.2 are checked against the Population and Business Registers respectively. Thereafter, the names are no longer needed in the Throughput and Production Databases. The home address and address of the establishment in Chart 2.2 can be used to improve the addresses in the Population and Business Registers respectively; but are thereafter replaced with regional codes in Chart 2.3. Job title and the employer's internal codes for occupation (TNS) are replaced with statistical ISCO codes for occupation. A derived variable for full-time salary is also created in Chart 2.3, and an educational code is imported from the Education Register.

Chart 2.3   Corresponding statistical register – processed data in throughput database

anPIN	Residential municipality	anBIN	anEsIN	Establishment municipality	Occupation ISCO	Education code	Actual salary	Extent of work	Full time salary
12345678	0180	654321	34567	0180	2222	1234567	18 342	0.60	30 570
23456789	1881	654321	45678	1880	3333	7654321	45 780	1.00	45 780

In this example, having access to the original job title and TNS-code would be advisable for any future changes in occupational classifications. The simplest method would be to take the administrative sources out of the data system and to archive them in a locked space once the statistical register has been edited and processed. This would make carrying out future controls possible without the administrative data being available in the data system.

Names, addresses and other details in plain language should only exist in the Input Databases. Only a small number of people within the statistical office should have access to these variables. The Throughput and Production Databases contain the statistical registers, and names, text, etc. should not be allowed there.

After this replacement of sensitive identifying variables, all databases with data regarding persons can be integrated with the anonymous number *anPIN*. Thus, the main part of the staff working with register-based statistics will not have access to names, official identification numbers, etc.

Communication variables

Names and residential addresses are necessary for data collection in sample surveys, but this information can be very sensitive. This information should only be stored in the Population Register at the NSO, and only a small number of persons at the NSO should have access to this information. When a sample survey of persons will be carried out, the *anPIN* numbers for the sample are sent to the Population Register. Only the names and addresses for the sample are delivered to the unit responsible for the sample survey. Sample surveys of enterprises or establishments should be handled in the same manner.

Combinations of identities

Several administrative registers contain combinations of identities. For example, in the public Vehicle Register, the car plate number and the car owner's official *PIN* are combined. If there is a data matrix at the NSO, where car plate numbers and the anonymous *anPIN* are combined, a person at the NSO can find the official *PIN* for car owners with the public Vehicle Register.

All official identities should be anonymised to prevent finding the official *PIN* related to an anonymous number *anPIN*. Therefore, vehicle registration numbers, address codes, real estate identities and Business Identity Numbers (*BIN* and *EsIN*) for enterprises and establishments should also be anonymised before they can be used for statistic production at the NSO.

Legislation against re-identification

There is always a possibility for a person with access to registers to re-identify microdata, even if the *PIN* has been replaced by *anPIN*, and the names, etc. have been removed. For example, it is possible to find the authors of this book if you know our home municipality, level of education and occupation. You will find our *anPIN* numbers and can thereafter search in all registers for information about us.

Because of the risk of re-identification, the NSO should not make anonymised registers available for analysis. Only random samples from registers without identities can be made public.

Legislation is the only protection against this kind of violation of privacy. According to Swedish law, this crime can be punished by imprisonment for one year. Every person who is employed by the statistical office must sign a document in which the security rules are clearly stated. Managers are responsible for employees fully understanding the importance of these rules.

2.3   Second obstacle: How to take care of dirty data?

Assume that an NSO has made agreements with some ministries and has gained access to several administrative registers. A long learning process starts for the statistical office – how should the new data be interpreted and how should the quality be analysed? For ministries and other authorities responsible for administrative systems, the new cooperation with statisticians will result in feedback regarding data quality. This feedback can start improvements in administrative systems.

Example: Criminal statistics in Peru

INEI, the statistical office in Peru, was asked to develop a new system for criminal statistics. Microdata with identity numbers of persons were created by many clerks working for two ministries, the police and the correctional authorities. The information was recorded through handwriting or with different IT systems and different authorities also used different codes for classification.

INEI's new system based on *one form* will standardise these kinds of reporting. The system has two modes:

– Recording of microdata into the system
– Retrieving statistical information (statistical tables) from the system

The system for criminal statistics focuses on number of crimes, number of victims and number of criminals. To enable comparisons, these numbers must be transformed into intensities showing number of crimes/victims/criminals as percentages of the corresponding population. The system handles this through comparisons with the previous census.

The lesson we can learn from this example is that instead of working with dirty data, bad administrative systems should be improved. This will benefit both the statistical office and the administrative authorities.

2.4 The new production process

How should the new data from the administrative registers be interpreted, and how should the input data quality be analysed? These challenges mean that the statistical office must develop a new production process. The NSO must have access to new kinds of metadata and develop methods for editing and quality assessment.

The *production process model* in Chart 2.4 describes how the work should be organised within the statistical office. We structure the process in four steps numbered 1–4 in the chart.

Chart 2.4 Transforming administrative registers into statistical registers[2]

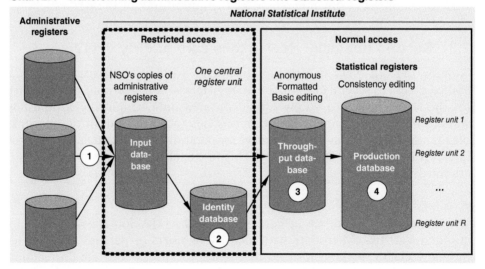

---

[2] This model was developed together with David Chaín, INEC in Ecuador.

1. Administrative registers are delivered by administrative authorities to the national statistical office. The selection of units and variables for submission must correspond with the statistical requirements. Cooperation between the administrative authorities and the statistical office should be well established and permanent. Only new and changed observations should be submitted for updating the Population and Business Registers.

2. With the Identity Database identifying variables, official identity numbers and names are replaced by anonymous numbers. Only these anonymous identity numbers are transmitted to the Throughput Database.

3. The variables in the Input Database are transformed and formatted so that all variables in the Throughput Database comply with the standards and formats used at the statistical office. In addition, *basic editing of the variables* in each administrative register is performed in the Throughput Database.

4. The main work of creating statistical registers is done in the Production Database. Different sources are combined, register populations are created, derived units and derived variables are created, and *consistency editing* is carried out.

Consistency editing is described in Chapter 6. Errors regarding coverage, units and variable values can be found by making comparisons with different registers and with sample surveys.

2.4.1   Contacts with administrative authorities

For the national statistical office, contacts with administrative authorities are necessary for many reasons. The statistical office needs not only microdata but also metadata with definitions of all administrative variables. The administrative process that generates the administrative registers must be understood – how has the administrative object set been generated? These object sets will be used to create statistical populations. Therefore, the statisticians must understand the administrative system in order to judge the relevance of the data.

The selection of units and variables submitted to the statistical office must correspond with the statistical requirements. The authority often has much more data than the statisticians expect. To avoid misunderstandings, the selection of data to be submitted must be decided after a thorough discussion between the two partners.

Cooperation with administrative authorities should be handled by working groups from the statistical office. These working groups should consist of staff with subject matter competence, competence regarding register-statistical methods and IT competence. Cooperation between statisticians and different ministries and administrative authorities should be a long-range project that aims to develop the national administrative and statistical systems.

2.4.2   Metadata have a new role

Documenting one of the NSO's own sample surveys is comparatively simple. The variables are defined by the questionnaire, and the frame and sampling methods are usually the same as in the previous version of the survey. The documentation is made primarily for future *users* of survey estimates. Colleagues working with other surveys have little use of the documentation in their work.

When an NSO gains access to administrative registers, the first thing the staff needs is metadata. How should the new data sources be interpreted? There should be one documentation of each administrative register with a definition of the administrative object set and definitions of all administrative variables. The metadata should be formalised and stored in electronic form in the NSO's metadata system. This documentation is made for the *producers* of register-based statistics. The producers need the metadata to judge the usability of an administrative register.

Microdata from one sample survey are for one specific use only; but one administrative register can be used in many register applications. Therefore, the metadata should be available to the entire staff. Different specialists with different subject matter knowledge should look at new administrative registers and check it there are some variables or some subsets they can use.

Example: Register of the disabled

This administrative register could be used for different purposes:

  – For producing statistics on disabled persons.
  – To improve residential addresses in the Population Register.
  – Economic support to the disabled could be included in the Income Register.
  – The health of the disabled could be included in a general health register.

A metadata system for input data from administrative registers could be stored in a separate system, so that it is not confused with the metadata that is generated for the final statistical registers. This system should include the following parts:

Chart 2.5   Metadata system for input data from administrative registers

From the administrative authority's metadata system:	From interviews with the administrative authority staff:
Definitions of object set and variables Stored as formalised metadata	How does their administrative system work? What do they know about the quality? Notes from discussions stored in text-files

From the administrative authority:	From the NSO's work with editing each administrative register:
All administrative forms, brochures, and instructions, in paper form and electronically	Quality indicators describing the most important types of errors. Stored as time series

2.4.3   Anonymisation of identity numbers

All identity numbers are handled in the Identity database in Chart 2.4, and the official numbers are replaced by anonymous numbers. This database should be the only place in the IT system (and in the country) where the links between the official identities and the anonymous identity numbers are stored. No copies should be kept outside the NSO.

In Chart 2.6, this part of the system is illustrated by an example regarding *PIN*-numbers that could be used in Sweden. Sweden has a population of about 10 million persons, which implies that eight digits are needed for the anonymous numbers. We use numeric format for these anonymous numbers to make IT processing as easy as possible.

Chart 2.6   Identity database

row	PIN	anPIN
1	19161234-6105	42393827
2	19161234-8720	48127259
3	19161234-9200	37873567
4	19161235-1783	31709824
5	19161235-4563	56142801
6	19161235-5651	66550635
...	...	...
9889184	20150132-6834	38177615
9889185	20150139-7594	90710026
9889186		55667435
9889187		73858794

We assume that there are 9889185 persons in the database when we start. The last person in the database table has *PIN* 20150139-7594.

In the *anPIN* column we store a random permutation of the numbers 10000001–99999999.

The next baby born is assigned a *PIN* at the hospital and this *PIN*-number will be linked with *anPIN* 55667435.

The next person, baby or immigrant, will be linked with the *anPIN* number 73858794.

2.4.4   Editing of a single administrative register

The editing of a single administrative register should consist of three components: first, standardising of variable names and variable formats; secondly, the checking of the object set so that there are no duplicates; and thirdly, the same kind of editing that is common in sample surveys – to find and correct erroneous variable values.

Standardising formats and names

When an NSO starts receiving administrative registers from different ministries and register-keeping authorities, the statisticians will find that the same variable has different formats and different names in different sources. One ministry can name a variable as *sex*, with codes *0* = *men* and *1* = *women*. Another ministry can name the variable as *gender*, with codes *1* = *men* and *2* = *women*. When input data are transferred from the Input Database to the Throughput Database, both variable names and formats must be standardised and follow the standard used at the NSO. The ideal would be that all authorities change to the NSO standard.

Checking duplicates

Some administrative sources can contain duplicates. Finding and deleting these duplicates are important; otherwise, matching of registers with duplicates will generate more duplicates. If two registers, each with one hundred duplicates, are matched, then the combined register can include $100^2 = 10\ 000$ duplicates.

Example: A Population Register was checked for duplicates in the following way. The register contained identity numbers for some persons and names, birth dates and birth places for all. Some 11% of the register's object set had no identity number. The identity numbers were checked for duplicates and about 0.2% of the observations had the same identity number as another observation. These duplicates in identity numbers were checked. In some cases, different persons had the same identity numbers, and in other cases the same person had two observations but there were errors in the spelling of names or in the date of birth.

When identity variables can have errors, then *probabilistic record linkage* (discussed in Chapter 4) using many identifying variables as matching key is a method for finding potential duplicates.

Checking variable values

This section discusses Statistics Sweden's register of *Yearly Income Tax Returns from enterprises* as an example of editing register data. There are different tax returns for sole proprietorships, trading partnerships, limited partnerships, limited companies and economic associations. Yearly Income Tax Returns consists of three parts: balance sheet, profit and loss statement and tax adjustments. Limited companies provide more information and sole proprietorships give less.

The register of Yearly Income Tax Returns is the main source for the Structural Business Statistics (SBS) survey, and a micro-simulation model used by the Ministry of Finance. The Yearly Income Tax Returns is also used by other surveys at Statistics Sweden. The data with tax returns consist of data based on four different tax forms for different kinds of enterprises or legal units. These four kinds of tax return data are edited separately. Simple subject-matter-based *automatic editing* software has been developed for each kind of tax form.

Some results are presented here from the editing of tax forms from limited companies. The total number of tax forms from limited companies for a specific year was 368 432. Of these, 16 571 were detected and corrected by the editing software as containing at least one error. The statistically important variables are in the balance sheet and the profit and loss statement. Of the 368 432 observations, the editing software found and corrected 693 tax returns with at least one error in the balance sheet. In the profit and loss statements, the editing software found 619 tax returns with at least one error. Turnover and three variables with costs for goods and services (Costs A, B and C in Chart 2.7) have the largest errors. The table below describes the errors found in these four variables.

Chart 2.7   Errors in profit and loss statements from limited companies, SEK million

Turnover		Costs A		Costs B		Costs C	
3 errors:	Corrected	4 errors:	Corrected	1 error:	Corrected	16 errors:	Corrected
0.989	0.983	0.002	−0.002	0.005	−0.005	0.620	−0.620
716 716.174	0.716	0.646	−0.646			0.062	−0.062
0.151	0.158	−5.140	5.140			58 035.580	0.580
		−0.048	0.048			0.018	−0.018
						2.092	−2.092
						...	...
						0.191	−0.191
						33 434.093	0.343
						0.585	−0.585
						0.020	−0.020

Among the 368 432 *turnover* values, only three errors were found, two small errors and one technical error with large impact. A small number of sign errors and two technical errors were found in the three variables in the *Costs for goods and services* group. Sign errors and large technical errors are easy to find and correct.

Finding and correcting these large technical errors are important; otherwise, a fatal error can be made somewhere among all those who use the data. Hence, a central unit receiving these data must perform this kind of editing before others gain access to the data at the statistical office. That is why this editing should be done in the Throughput Database by a central team before it is used by many different teams in the Production Database. In Chart 2.8, the profit and loss variables are summed for all observations in the register with data regarding limited companies.

Chart 2.8   Profit and loss statements for limited companies, SEK million

	Before editing	After editing	Difference
1. Turnover	7 007 567	6 290 851	716 715
2. Other receipts	110 593	110 593	0
3. Costs for goods and services	−4 672 119	−4 580 652	−91 468
4. Costs for labour	−1 127 633	−1 127 622	−11
5. Value added (1 + 2 + 3 + 4)	1 318 407	693 170	625 237
6. Depreciations	−360 379	−359 763	−616
7. Financial costs/returns	279 157	279 160	−3
8. Tax	−80 183	−80 166	−17
9. Profit	533 949	532 277	1 673
Inconsistency (5 + 6 + 7 + 8−9)	623 052	125	

The errors in this administrative source have serious effects. But the number of observations that have errors is small and they are easy to find and correct. Some errors remain in the corrected data (125 SEK million). These errors can be detected and corrected in subsequent, more advanced, editing.

What can be learned from this example?

First, it must be remembered that we are editing *administrative* data, not statistical data based on a questionnaire created at the statistical office. This requires the development of subject-matter competence at the statistical office. In this case, complicated tax forms and taxation rules must be understood by the staff who use these data for statistical purposes. Regular contacts with staff at the National Tax Agency are necessary to develop and maintain this competence.

We also observe that data from the National Tax Agency can contain large errors. The National Tax Agency processes one tax form at a time and looks at the entire tax form for each enterprise. Errors are noticed and understood, but the database is not corrected. A taxation decision is made based on each tax form, where the large technical errors we have seen in the example above have been eliminated. Therefore, the statistical office should also obtain these *taxation decisions* and use that information for editing purposes.

Finally, information about the character of the administrative data should be used when the editing methods are developed. When editing data from tax forms, we should make use of the fact that some variables are legally important. The monetary sums that are the bases for taxation must be correct; otherwise, there is a risk of prosecution. If these *go-to-jail* variables can be assumed to be correct, it will be easier to understand where the errors are.

2.4.5   Organising the work with administrative registers

The Identity Database is developed for two purposes – to protect integrity and facilitate register processing. Integrity will be protected as names and official identity numbers will be removed – the staff working with register-based statistics will only have access to anonymous identity numbers when they create their statistical registers.

However, there can be parallel systems of identity numbers and each system should be linked to the *anPIN* numbers. One specific person can have two id numbers: one taxpayer number and one health system number, for example. Both numbers should be linked to the same *anPIN* number. The Identity Database will then make the production process easier and reduce the risk of mistakes. This is because all data used in the Production Database have been edited so that complicated identifying variables as names and different kinds of identity numbers have been replaced by a unified system of anonymous identity numbers that are easy to use.

A central unit with a limited number of staff should be organised as responsible for receiving all administrative registers. Only these persons should have access to the servers and databases where copies of the administrative registers are stored. The central register unit is also responsible for replacing names and official identity numbers with anonymous identity numbers. Only the central unit, therefore, has access to the Identity Database where the links between identities and anonymous identity numbers are stored.

Do it right from the beginning!
The Identity Database and the standards and formats that should be used for the variables in the system should be established early in the process of modernising the statistical system. If this rule is not followed, expensive IT systems will be developed that use original identities and formats in the administrative sources. Changing these IT systems will be quite costly when people later understand that privacy must have better protection and that standardisation is necessary.

The Production Database
The statistical registers are created by different register units in the Production Database. Each unit imports the administrative registers it needs from the Throughput Database. The unit can also use the statistical registers in the system – all statistical registers in the system are stored in the Production Database. The only identity of persons is the anonymous identity number. Statistical populations, units and variables are created by the responsible register units. These populations, units and variables are then checked by consistency editing. This editing work results in the discovery and reduction of errors and inconsistencies.

The editing work in the Throughput and Production Databases is the main part of the work with quality assurance of register-based statistics. Documentation of the results of the editing work is important – this will make it possible to monitor how the quality of administrative data changes over time.

2.5   Third obstacle: The national registration system

The term *national registration system* refers to all the administrative systems in the public sector that are used to handle the national registration of persons living in the country and their activities. These activities include, for example, studies, jobs, tax reporting and tax payments, health and social services and participation in elections. Registration of enterprises and their activities are also included in the national registration system and also ownership of real estate and vehicles. Identities of taxpayers, students, patients and owners are as a rule combined with the administrative transactions.

The NSOs want to use this gold mine of information for statistical purposes, but there may be laws and regulations that prevent the NSOs from accessing the data. In addition, there may be quality problems that prevent register data from being useful for statistics production.

In this section we describe the problems regarding legislation and quality that we have seen in Latin America and the Caribbean. Bycroft (2015) discusses similar problems in New Zealand. The NSOs need a twofold strategy for these problems:

- Improve preconditions in the long term. This must be done, but will take time.
- In the short term, develop methods to use the available registers today.

2.5.1 Legislation governs access to data

Legislation does not allow data sharing in several countries in Latin America, the Caribbean and in New Zealand – ministries and national offices are not allowed to cooperate and share microdata. There are registers with births and deaths, but due to privacy concerns some countries have no central Population Register and no unique identity numbers for persons.

However, in all these countries there are registers with taxpayers with tax identity numbers; the national health system has patient registers with patient identity numbers; the national education system has student registers with student identities, etc. The national identity card may contain several different identity numbers – the ministries do not cooperate, but these systems of identity numbers could be linked.

Several countries that not yet have given their NSOs access to administrative registers have unique national identity numbers. When the ministries in one of these countries decide that the country can benefit by combining data from different ministries for some important purpose, such as support of poor people, then microdata can be integrated. There is legislation to protect privacy, but the public benefits of integration outweigh privacy concerns about the use of integrated microdata.

The NSO could join these projects with data sharing. The public benefits from replacing the traditional census by integrating registers is a project that yields great benefits to the country. Formal legislation is not the whole problem; attitudes that influence the interpretation of existing legislation are also important.

2.5.2 Too many registers, but no good registers – what to do?

In one country, seven different national offices handle registration of the population of persons and the demographic events that affect that population. In Sweden and the other Nordic countries, there is only one office that handles registration of persons and the demographic events regarding persons.

Chart 2.9   Registration of persons & demographic events in a Latin American country

Population registers	Civic registration	1 Institution
	National identity cards	1 Institution
Demographic events	Births	2 Institutions
	Migration within the country	0 Institutions
	International migration	1 Institution
	Deaths	2 Institutions
In all		**7 Institutions**

Duplication of work costs money, but it also results in inconsistencies. Sex, birthdate, birthplace and home address are variables that appear in all registers created by these seven institutions. Variable values can be old and outdated in some registers. Spelling errors and errors in registration will occur. If these institutions could work together, these errors and inconsistencies could be reduced.

In Chart 2.9, all institutions work with the same kind of data, the demography of the country. In Chart 2.10, the different ministries create registers that share some basic variables, but most variables differ. But here also, there is duplication of work and inconsistencies regarding the basic variables for each person. Moreover, the object sets of persons differ significantly – the Ministry of Health has only patients and persons who have been in contact with health institutions in their registers, the Ministry of Education has only pupils and students, etc.

As a consequence, for example, the Ministry of Health cannot compute fertility rates by age group and region, because all women are not in their registers. And for the missing women, they have no information regarding where these women live.

Chart 2.10   Uncoordinated work with registers

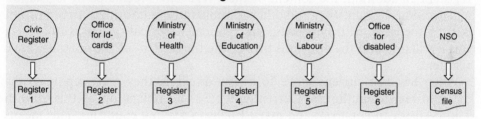

In Chart 2.10, we have many registers, but no good registers – what to do? *Cooperation* is the answer to this question. Chart 2.11 illustrates the Nordic way of working with registers. *One* national authority is given the responsibility to create *the* central Population Register in the country. All persons living in the country should be registered and all demographic events regarding persons should be registered by this national register authority. All other ministries and national offices that need access to population data use the central Population Register. And *these users are mandated* to report all errors they find to the national register authority.[3]

---

[3] There is one exception: Due to the *principle of one-way traffic*, the NSO will not give any information regarding persons with erroneous values in the central register.

Chart 2.11   Cooperation regarding the central population register

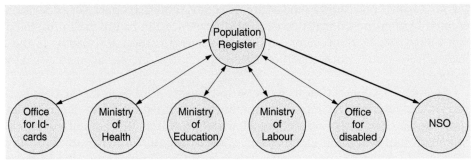

The quality of an administrative register depends on whether the register is used often or seldom. Users will find errors and if this information is used to correct the central register, the quality will gradually improve. Then all users of the central register will receive the same basic information, which will be continuously updated and improved. Then users can add their own variables.

2.5.3   Legislation rules obligations to report and what to report

Legislation is not only important to give the NSO access to administrative registers. In addition, the content of these registers is regulated by legislation and decisions by ministries and other administrative authorities. An important part of the work with improving the national statistical system is to change this legislation and the decisions that define the content of the administrative registers.

When Swedish citizens move to a new dwelling, they must report to the national register authority. Currently, there is no obligation of this kind in Latin American countries and many English-speaking countries. This means that preconditions for regional population statistics differ considerably between countries. If a country wants to develop register-based regional statistics, legislation will be necessary so that notifying an authority is mandatory when people move to a new dwelling.

Swedish tax returns are quite easy to handle for most people. The well-established national identity numbers and the central Population Register are used by employers, banks, pension funds, the social insurance agency and others. These actors transfer most of the required data to the tax authorities without involving the taxpayers. This improves the quality of the information in the tax forms. If a country wants to improve preconditions for register-based social statistics, then a central Population Register and a national identity number for persons should be developed.

The variables defined on different kinds of tax forms and other administrative forms will be the bases for the statistical variables in the register surveys developed by the NSO. As these variables are for combined use – both tax administration and official statistics – the NSO should work together with the ministries and administrative authorities to improve administrative forms.

The NSO can contribute its expertise regarding design of questionnaires and make the administrative forms easier for the respondents. The NSO also knows what information is needed for official statistics – for example, how do the tax forms for enterprises comply with the needs of the National Accounts?

2.6   Why has the census been so important?

Where do you live, and with whom? Where do you work? These three questions are perhaps the first things we want to know about people we meet. The traditional census answers these questions and provides small area estimates regarding, among other things, population, households and employment.

In most countries without register-based statistics, the census is the only available source that can give this kind of information. Some countries spend substantial amounts to replace the census with a yearly sample survey, where many persons must answer census-like questions. However, many countries plan to replace the traditional census with a completely or partially register-based census due to rising costs and increasing quality problems with the traditional census.

Where do you live, and with whom? Where do you work? These questions and other questions in the traditional census are answered by different statistical registers.

Let us return to the following chart from Chapter 1 that shows the statistical registers developed by all the Nordic countries.

The Population, Employment, Education, Income & Taxation, Real Estate and *Dwelling Registers* are the yearly registers that replace the traditional population and housing census. The creation of these registers also requires the use of the Activity and Business registers.

From Chart 1.2   The Statistical System in a country can consist of the following surveys:

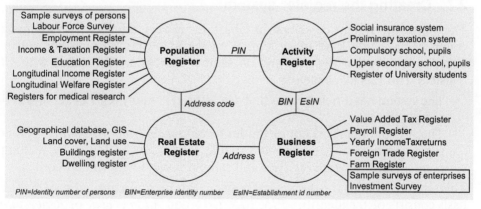

In the traditional census, all variables are collected together. The census tables can be based on any combination of these variables, and all estimates in all tables will be consistent. When the census has been replaced with the statistical registers in Chart 1.2, all variables in the registers can be combined with the identity numbers and address codes.

The four registers in Chart 1.14 are created at different points in time. The register-processing must be coordinated to obtain consistent estimates from different registers. First, the team responsible for the Population Register creates a specific register version regarding a specific point in time, usually 31 December. Thereafter, the other registers create their employment, education and income variables for exactly that population. Note that these registers have information regarding more persons than the persons included in this specific population.

From Chart 1.14   Register-based census statistics for one small municipality

Age	Population Register — Number of persons	Employment Register — Em-ployed	Not em-ployed	Com-pulsory	Upper secon-dary	Post secon-dary	Post-graduate	Not known	Income Register — Yearly earned, SEK thousands (1 USD ≈ 8 SEK) 0	1-139	140-279	280-
0–15	1 358	-	-	-	-	-	-	-	-	-	-	-
16–19	384	84	300	297	67	0	0	20	120	259	5	0
20–24	328	197	131	55	233	31	0	9	38	153	105	32
25–34	719	622	97	88	423	201	1	6	15	141	320	243
35–44	962	846	116	83	554	319	2	4	19	106	381	456
45–54	910	776	134	171	500	226	4	9	24	89	363	434
55–64	1 071	791	280	295	491	271	7	7	21	145	468	437
65 +	1 402	138	1 264	758	438	194	8	4	1	508	719	174

(Education Register spans: Compulsory, Upper secondary, Post secondary, Post-graduate, Not known. Income Register spans: 0, 1-139, 140-279, 280-)

2.7   Creating the register system

The statistical system described in Chart 1.2 is very advanced – not many countries have that kind of system. With the identities, *all* registers can be linked with *all* other registers and *all* sample surveys can be linked with *all* registers. This has two benefits:

- The system has a rich *content* that can be used in many register surveys and sample surveys.
- The system offers new opportunities to *analyse and improve quality*. Errors regarding coverage, units and variable values can be found by making comparisons between registers and sample surveys with similar content.

This section explains the theoretical basis for this kind of national system. The basis consists of the identities *PIN, BIN, EsIN* and *Address code* or *Real estate identity*. At least one of these identities must be present if an administrative register or a sample survey can be included in the system. The development of a completely register-based system also requires some administrative registers that include *combinations of identities* as well as administrative systems that regularly update these combinations.

2.7.1   Where do you live?

The transition towards a completely register-based system started in Sweden in 1967 when the central Population Register was created. At that time, this register included the official identity number for persons (*PIN*), and the official identity of the real estate where a person was registered as permanently living. The entire population of persons was georeferenced by this important combination of two identities. Then it was possible to produce regional population statistics for small geographical areas and small categories of persons. Other registers with *PINs* could also be included. The system at that time is illustrated in Chart 2.12.

Chart 2.12   Sweden's system 1967–1984

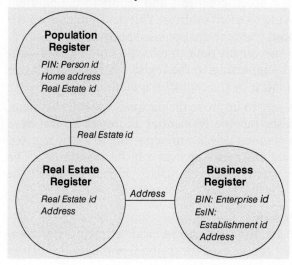

Persons can be linked to the real estate where they live. When people move to a new real estate, the link is updated.

A real estate can consist of buildings with many apartments and then it is impossible to know which persons live together in the same apartment and form a dwelling household.

This system has no link between persons and establishments, so it is not possible to answer the question where people work.

The central Population Register also contained the home address for all persons. These addresses were used by authorities for sending postal mail to citizens regarding taxation, elections, etc. But these addresses were not standardised in a way that made them suitable for record linkage – that is why real estate identities were used instead.

The Nordic countries have a long tradition of notifying the national register authority when a person moves to a new address. This reporting is mandatory, but as many social benefits are related with the municipality where you live, the quality of the home address in the Population Register is good with one exception: we estimate that about 16% of persons 20–24 years of age have the wrong municipality in the Population Register. Young people moving away from home for the first time can be a difficult category to georeference.

In addition, it is important that reporting the new address is easy. One report to the national register authority should be sufficient. Since all authorities use the central Population Register, all authorities, banks and insurance companies are automatically informed.

Population statistics in the long term

A long-term national strategy must be pursued with the following objectives:

- Developing and maintaining the central Population Register should be assigned to a national register authority. All demographic events: births, deaths, external migration, internal migration and changes in martial status will be handled by this central Population Register.

- A unique personal identity number handled by the national register authority should be used by central and local administrations. Newborn and new immigrants are given identity numbers from the national register authority.

- All persons permanently living in the country should be required to register as permanently living at a specific address and report to the national register authority when moving to a new address. This new mandate on the citizenry should be introduced gently so that public approval is developed. Central governments often use census data to allocate funding. This fact could be used to explain why reporting to the register is important when the population census is replaced by the central Population Register.

Ecuador has been working this way to improve the national statistical system. A civic Population Register exists, but the residential addresses do not have sufficient quality. According to planned legislation, reporting to a new register, the *Register of Residence*, where each person lives will be mandatory. The dwelling is defined by the georeferenced electricity meter used by the household.

Population statistics in the short term

A brief overview follows of the methods that can be used to create a statistical Population Register. This register can be used for regional population statistics until a central Population Register has been developed.

A statistical Population Register can be created by the following measures:

- An inventory of relevant administrative registers should be done. Two kinds of sources are needed:

  1. Sources that can improve coverage of the statistical Population Register.

  2. Sources that can improve the quality of residential addresses.

- Existing household sample surveys should be used to measure the quality of the statistical Population Register. What percentage of the population lives in the region, municipality, district or dwelling that is listed in the statistical register? The sample surveys should include personal identity numbers to facilitate record linkage with the statistical register.

- If the quality of microdata is considered insufficient, more administrative sources should be used. Estimation methods can also be developed that reduce errors in the population estimates. This is discussed in Chapter 8.

2.7.2   Where do you work?

What should those responsible do so that register-based employment statistics can be produced? In Sweden, no administrative registers with data regarding establishments existed before 1985. The enterprises had to deliver tax reports and pay social security and preliminary taxes for their employees. The enterprises are the legal units, and only legal units had these legal obligations.

The decision was taken to move a step closer to a completely register-based census by creating a new tax register from 1985 onwards. The register should consist of georeferenced microdata regarding annual pay for each job. The unit *job* is a combination of two units: *employee* or *sole proprietary* with identity number *PIN* and *establishment* with identity number *EsIN*. This created no problem for enterprises with only one establishment. However, for enterprises with more than one establishment, the enterprises now must specify for each employee at which establishment the employee worked.

Chart 2.13   Sweden's system 1985–2010

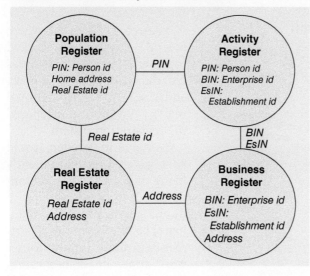

Since 1985, the register with all job activities contains combinations of the personal identity number *PIN* and the establishment identity number *EsIN*.

Job activities constitute a part of the Activity Register together with study activities. Statistics Sweden is responsible for the Establishment Register, which is a part of the Business Register. The Establishment Register contains addresses of the establishments.

The four registers in Chart 2.13 are from 1985 and onwards. These are the *base registers* in the Swedish statistical system. They define important populations and link different kinds of units. Both persons and jobs are now georeferenced.

This register system can now create regional employment statistics. Persons in the Population Register can be linked with their jobs in the Activity Register. These jobs can be linked to the establishment in the Business Register, and each establishment can be georeferenced through the address. Since persons can be linked to the place where they live, register-based statistics on commuting can also be created.

To be able to produce these statistics, the Employment Register was created based on tax and social security data regarding job activities. This required support from politicians and cooperation with the Swedish Tax Agency. All employers with more than one establishment must deliver statistical information regarding where each employee worked combined with the reporting for taxes and social security.

2.7.3   With whom do you live?

The residential addresses in the register should specify the *dwelling* where each person lives. Using these dwelling identities, it is possible to create dwelling households. Real estate identities and residential addresses were the starting point for the work to improve Sweden's statistical system regarding dwelling identities. However, the residential addresses did not have a standardised format and could therefore not be transformed into numerical codes. There were registers that defined real estate – building – entrance – floor; but there was no register with the apartments or dwellings on each floor.

Chart 2.14   Sweden's system 2011

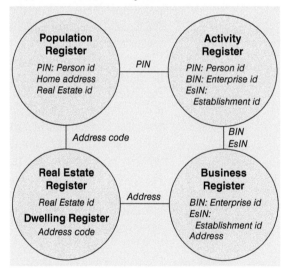

A project was started where several authorities worked together. The municipalities did the field work to correct addresses and the national register authority prepared the registration system.

Everyone in Sweden would be registered as permanently living at a specified and standardised address and in a dwelling with a specified number. The numbering of apartments in multi-family houses was done in collaboration with the property owners and landlords.

In the Nordic countries, dwelling registers were created in this way. All persons in the population were registered as permanently living in a specific dwelling. Dwellings can also be defined with georeferenced electricity meters or water meters. The population is then registered with the identity number of the meter connected to each household. This is a solution planned to be used in Ecuador. Other countries in Latin America will probably use electricity meters in this way.

2.7.4   A centralised or decentralised national system?

The transition from a traditional statistical system into a register-based system will also change the role of the national statistical office. In the new system, ministries and national offices must cooperate and share data. Coordination and common standards will be necessary, and the NSO should have the leadership in the transition work. Ministries, the central bank and other national offices should accept that the NSO becomes a stronger actor and that the statistical system becomes more centralised.

The system of base registers

There are methodological reasons why a more centralised statistical system is preferable. The four registers in Chart 2.14 are of fundamental importance; the links between these registers are between different object types. The link between *person – dwelling* in the Population Register and the link between *person – establishment* in the Activity Register are necessary for regional statistics and when the traditional census is replaced with statistical registers. For regional labour market statistics all four registers are needed. Because of the importance of these four registers, we have defined them as the *statistical base registers* in a register-based national statistical system.

These four statistical base registers should be developed and maintained by one actor in the statistical system, the NSO. The NSO should monitor coverage and consistency between the statistical base registers:

– All dwellings in the Population Register must also be in the Dwelling Register.
– All persons in the Activity Register must also be in the Population Register.
– All enterprises and establishments in the Activity Register must also be in the Business Register.

Activity Register – Population Register

According to the Swedish Labour Force Survey (LFS) for December 2018, there were 350 900 employed persons in construction. The LFS is based on the Population Register with persons permanently living in Sweden. According to a register created by the Swedish Work Environment Authority, there were 14 040 persons in construction temporarily working for a foreign employer in Sweden at 1 December 2018. There were also persons temporarily living in Sweden, working in construction for a Swedish employer. These two categories are missed by the Swedish LFS.

Activity Register – Business Register

According to Statistics Sweden's Business Register used for yearly statistics for a specific year, there were 57 286 enterprises registered as '*active employers*', which had not paid any kind of preliminary tax or social security for that year. In addition, 34 574 enterprises were not included in the Business Register, but these had paid preliminary tax and social security as active employers for that year according to the Activity Register. Some 274 232 enterprises were active in both registers.

These two examples show that the heart of the national statistical system, i.e. the base registers in Chart 2.14, must be monitored for coverage errors and inconsistencies regarding populations. To do that, the NSO must be responsible for the four statistical base registers (that can differ from the administrative versions).

Census statistics and statistics for the National Accounts

All registers created to replace the traditional census must be consistent. This is discussed in Section 2.6. The register processing should be coordinated to achieve this consistency.

The work with National Accounts will benefit from consistency between tables delivered by different surveys. These surveys can be linked and coordinated with the registers that are involved.

The coordination noted here requires that the NSO should be responsible for the registers that will replace the census or are used by the National Accounts. We consider that the NSO should be responsible for the registers in Chart 1.12 in Sections 1.1 and 2.6.

2.8   Register surveys and estimation methods

Perhaps most statisticians do not associate this chapter with survey design or estimation methods. However, survey design and estimation methods aim to achieve good quality estimates and use resources efficiently.

Section 1.1 discusses the term *survey system design*, which refers to the work of improving a system of surveys. We summarise the estimation methods discussed in Chapter 2:

1. Use unique personal identity numbers in all central and local administrations.
2. Develop a central Population Register that is used by all authorities. This register should have good coverage and residential addresses should be updated when people move to a new address.
3. Develop a dwelling identity (or electricity meter identity) to define households.
4. Develop an administrative system for job activities where persons are linked to the establishment where they work.

If this can be done, then the national statistical system in your country can answer the following questions with estimates from statistical registers: *Where do people live, and with whom? Where do people work?*

2.9   A traditional census or a register-based census?

A statistics producer such as Statistics Sweden can choose from the following survey methods for producing social statistics:

1. A traditional census, where interviewers collect data from all households with an area frame.
2. A sample survey based on an area frame, where interviewers collect data from households with face-to face interviews.
3. A register-based census where the NSO sends questionnaires to all respondents with a register that has good quality address information. The Swedish Farm Census has used this method. The last Swedish Population and Housing Census 1990 used this method with questionnaires with pre-printed identity numbers sent to all persons and households.
4. Yearly census-like tables are produced with a number of register surveys or statistical registers that the NSO has created.
5. A sample survey based on a frame created with a register, where the NSO collects data via telephone interviews or postal questionnaires.

Of the five survey methods above, only the last two, 4 and 5, have survived and are currently used by Statistics Sweden. This is due to the costs of the survey methods and the quality of survey data.

Costs and quality comparisons
Why should we use administrative registers for statistical purposes? There are two reasons for changing the national statistical system in this way: to reduce costs and improve quality. We always want to achieve these two goals when we work with survey design.

Chart 2.15   Survey design: reduce costs and/or improve quality

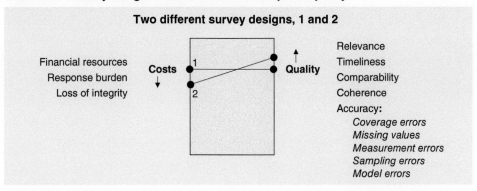

Chart 2.15 compares two survey designs. Assume that design 1 is a labour force survey in a traditional system and design 2 is a labour force survey in a register-based system. Then costs can be saved as mail questionnaires or telephone interviews can be used in a register-based system instead of sending out interviewers to visit an area sample of households. In addition, quality can be improved as sampling errors and nonresponse errors will be reduced when auxiliary variables in the register system are used to improve estimation.

If survey design 1 is a traditional population and housing census and design 2 is a register-based census, we can compare costs and quality in a similar way. The difference in quality is perhaps best illustrated by Charts 1.3 and 1.4 with employment for a small Swedish municipality – timeliness differs considerably.

Three kinds of costs should be minimised. *Financial resources* can be saved, and the *response burden* will be reduced by the transition to a register-based system. Traditional data collection will disturb the respondents in many cases. When people get a telephone call from the statistical institute during the evening, this can be experienced as a *loss of integrity*. After an interview, the interviewer may have obtained confidential information that could be a threat to the respondents' integrity. We can reduce these costs with register surveys.

In a register-based system, anonymous numbers to protect integrity should replace identities, and access to microdata in the registers should be restricted. If these measures are not taken, costs as losses of integrity may increase when a register-based system is developed.

When discussing quality, remember that a register-based system opens new possibilities. Instead of an expensive traditional population and housing census, you will have a number of register surveys that will provide yearly statistics that did not exist before. For example, yearly statistics for small regions will only be possible in a register-based production system. It will also be possible to create longitudinal registers that will be important for researchers. In these cases, we compare a situation with no estimates at all in the traditional system with the quality of the estimates from the new register surveys.

CHAPTER 3

The Nature of Administrative Data

The transition from a traditional census-based system to a register-based system implies that the National Statistical Office will use new kinds of microdata. Terms, methods and modes of reasoning from work with sample survey data are not always suitable when the NSO staff starts working with register data. Sampling errors lose their dominance, and measurement errors that are an important problem in household sample surveys may not be relevant for some kinds of administrative data. On the other hand, there are other types of errors that can become important problems when working with register data.

Different terms are used for different kinds of data. Section 3.1 compares *questionnaire* data and *register* data. Questionnaire data have been collected by interviewers or via mail sent to the respondents. Administrative data have been generated by an administrative system. These two kinds of data may look similar. In both cases, a form of one or two pages with questions must be filled in; nevertheless, the character of these two kinds of data can be quite different.

Section 3.2 compares *administrative* and *statistical* data. The two terms are used to distinguish between different ways to use the same data. Administrative data are data used to administer *individual objects*. Statistical data are data used to produce estimates for *aggregates of units*. Administrative registers at manufacturing enterprises are used as a general example to illustrate these two ways of using the same data.

3.1 Comparing questionnaire and register data

This section illustrates important differences between data based on statistical questionnaires and data based on administrative registers. These examples also illustrate the quality principle in Chart 1.15, which emphasises quality assessment based on microdata comparisons between different surveys.

3.1.1 A questionnaire to persons compared with register data

Statistics Sweden conducted a sample survey on commission using a postal questionnaire in January–April 2004. Two questions in the survey can be compared with similar variables in the frame – the current version of the Population

Register-based Statistics: Registers and the National Statistical System, Third Edition. Anders Wallgren and Britt Wallgren.
© 2022 John Wiley & Sons Ltd. Published 2022 by John Wiley & Sons Ltd.

Register at 31 October 2003. The following two questions were included in the questionnaire and are compared with similar register-based variables:

What is your yearly income before tax?				Do you have children and how old?		
– 79 999	☐	300 000 – 399 999	☐	Years:	Yes	No
80 000 – 149 999	☐	400 000 – 499 999	☐	0–6 years	☐	☐
150 000 – 224 999	☐	500 000 – 599 999	☐	7–12 years	☐	☐
225 000 – 299 999	☐	600 000 –	☐	13–17 years	☐	☐

In the grey cells below, the respondents' two income variables are consistent; and in the other cells, the two income variables are inconsistent:

Yearly income in questionnaire	Yearly income before tax according to administrative data, SEK thousands									
	Missing in register	-79	80-149	150-224	225-299	300-399	400-499	500-599	600-	All
-79	3	287	51	8	3	1	0	0	0	353
80-149	2	40	170	79	6	3	1	0	1	302
150-224	6	16	49	311	117	14	2	2	2	519
225-299	1	8	14	80	338	67	6	1	5	520
300-399	3	2	3	8	34	153	21	2	2	228
400-499	0	3	1	3	1	12	47	10	1	78
500-599	0	2	0	1	0	0	4	12	6	25
600-	0	0	2	0	0	1	0	2	29	34
Response set	15	358	290	490	499	251	81	29	46	2 059
Item nonresponse	6	55	33	28	11	10	1	0	1	145
Object nonresponse	31	473	339	483	357	165	51	30	25	1 954
All in original sample	**52**	**886**	**662**	**1 001**	**867**	**426**	**133**	**59**	**72**	**4 158**
Item nonresponse rate		6%	5%	3%	1%	2%	1%	0%	1%	3%
Object nonresponse rate		53%	51%	48%	41%	39%	38%	51%	35%	47%

The income variable
Persons with low income may be missing in the Income Register because they are not required to report income (15 cases).

Item nonresponse for the income question is selective; item nonresponse rates are higher in the two lowest income classes. Object nonresponse is also selective.

When microdata from the sample are compared with register data, we learn important lessons regarding the quality of the sample survey. Here, the nonresponse in the sample survey implies that the estimated income distribution is biased.

What is your yearly income before tax?
The main weakness of the sample survey is that the income variable is not defined:
- What income year: the previous year 2003 or the current year 2004?
- What kind of income: disposable income, taxable income, earned income or income including unearned income?

When Swedish taxpayers start to prepare their income self assessments, they receive a brochure (32 pages) that explains many kinds of income. Including precise income definitions with explanations in the questionnaire would make the questionnaire too burdensome and complicated for respondents and would increase the nonresponse rate.

The income variable from the Income Register is clearly defined: earned income including unearned income according to the income self-assessment for the calendar year 2002.

Do you have children?

This seemingly simple question is actually not simple at all. Again, the questionnaire does not define which children should be included: biological children, adopted children, foster children, children living in the same household, etc.

Assume that two persons divorce and after a while start new families with new spouses. If one of the new families now consists of two adults where one spouse has two children from an earlier family (who spend every second weekend in the new family) and the other spouse has one child from an earlier family (who now is taken care of by the previous spouse), how many children do these adults have?

There is information in the Population Register on biological and adopted children and children that live in the same dwelling as the respondent. With register data, the choice of one clear definition among a number of possible definitions is possible.

Children 0–6 years	Register data		
Questionnaire data	Yes	No	All
Item nonresponse	5	362	367
Yes	310	39	349
No	2	1 486	1 488
All	317	1 887	2 204

The grey cells in the table are inconsistencies.

Two persons with children 0–6 years according to the Population Register answered that they have no children 0–6 years.

39 persons with no children 0–6 years answered that they have children 0–6 years.

Children 7–12 years	Register data		
Questionnaire data	Yes	No	All
Item nonresponse	28	403	431
Yes	267	34	301
No	8	1 464	1 472
All	303	1 901	2 204

The same kind of inconsistencies is found for children 7–12 years and 13–16 years.

Children 13–16 years	Register data		
Questionnaire data	Yes	No	All
Item nonresponse	29	402	431
Yes	244	37	281
No	5	1 487	1 492
All	278	1 926	2 204

Conclusion – do you know what the answers mean?

In a sample survey or census, the researchers can formulate the questions just as they want, *but how should the answers be interpreted?*

In a register survey, the answers are precisely defined, *but are the variables relevant for statistical purposes?*

This example highlights an important difference between questionnaire data that have been collected during a sample survey or census and administrative data that have been generated by an administrative system. Many administrative variables are precisely defined by administrative rules and legislation, and those who provide the data must follow these rules. In a sample survey, the answers depend on how the respondents interpret the questions. Our knowledge of such cognitive processes is, as a rule, limited.

3.1.2   An enterprise questionnaire compared with register data

This example compares information on turnover from three different sources: one questionnaire used in Statistics Sweden's Structural Business Statistics survey (SBS) and two different administrative sources based on tax forms.

The SBS survey was designed so that the non-financial enterprises in the Business Register were divided into two parts: part 1 where a questionnaire was sent to about 4 160 large enterprises; and part 2 where the yearly income tax return (YIT) was used for about 1 million legal units. The columns in Chart 3.1 are defined as:

BIN	=	Business identity number of each legal unit/enterprise
SBS	=	Turnover according to Statistics Sweden's questionnaire
YIT	=	Turnover according to the yearly income tax returns
VAT	=	Turnover according to 12 monthly VAT returns
Distance	=	$|SBS - YIT| + |SBS - VAT| + |YIT - VAT|$

The 4 160 observations have been sorted by distance. The enterprises 160001–160005 are the worst cases, where reported turnover in the three sources differ greatly. How should these values be interpreted?

One interpretation could be that we can only trust 'real statistical measurements' (SBS) and that administrative data (YIT and VAT) often have poor quality. However, YIT and VAT can both be perfect as administrative data. The differences can be explained by the administrative routines used by the enterprises.

Chart 3.2 shows that the differences are large for only a limited number of enterprises. If the 100

Chart 3.1   Yearly turnover for the same enterprises in three sources, USD million (transformed microdata)

BIN	SBS	YIT	VAT	Distance
160001	7 179	11 941	8 089	3 175
160002	2 954	0	0	1 969
160003	843	3 561	918	1 812
160004	5 514	2 888	2 895	1 751
160005	26	538	2 536	1 673
160006	2 301	0	0	1 534
160007	2 211	0	2 239	1 493
160008	1 316	1 316	0	877
160009	638	638	0	425
160010	456	0	435	304
160011	141	141	0	94
160012	113	0	127	85
160013	65	0	63	43
164159	34	34	34	0
164160	19	19	19	0

or 200 enterprises with the largest differences are checked, the reasons behind the differences between the three sources may be understood, and the rest of the 4 160 observations can be used as they are.

Since enterprises can own other enterprises, some enterprises use different identity numbers in their tax forms. YIT can be reported using one BIN, and VAT can be reported with another BIN. This must be the case for all enterprises in Chart 3.1 except for the two last ones with BIN 164159 and 164160. With the exception of these two, all enterprises report turnover for a group of units with different BIN.

Chart 3.2   Distance for each ordered observation

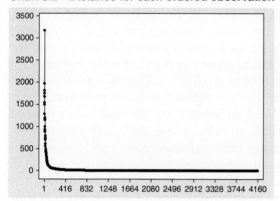

Chart 3.3   Complete groups of enterprises

BIN	SBS	YIT	VAT
160006	2 301	0	0
160666		2 301	2 301
160007	2 211	0	2 239
160777		2 211	0

Chart 3.3 shows two such groups. The first consists of the units 160006 and 160666. The SBS questionnaire was sent to 160006, but the enterprise group uses 160666 for YIT and VAT reporting. In the second group, the unit 160007 is used for VAT reporting and the unit 160777 for YIT reporting. The SBS values in both cases should have been 0, as turnover is reported for BIN 160666 and 160777. The latter belong to part two of the business population, where YIT is used as the data source. This shows the risk of double counting when a statistical questionnaire is combined with administrative data, as demonstrated in this SBS example.

Conclusion – do you know what the data mean?

In a sample survey, we must understand how the respondents react to our questionnaire and instructions. Similarly, we need to understand how those who report to an administrative system adjust their reporting to the rules and possibilities of the system.

In order to become well acquainted with the system, the NSO staff should study system metadata and have discussions with the staff at the administrative authority who handle the system. The NSO staff must also devote resources to analyse data carefully. An efficient way of developing knowledge about data quality involves making comparisons with other sources, as in this example.

In the turnover example, data that looked peculiar at the outset became meaningful if we combined the enterprises into groups. We also found measurement errors in the *questionnaire* part of the SBS survey. Measurement errors are not the main problem in the administrative sources here. The difficulty lies with the administrative units when we want to use the data for statistical purposes.

3.2   Enterprise registers for combined use

Enterprises and organisations have their own register systems that can also be used for administrative and statistical purposes. The accounting system at an enterprise is a good example of an information system that is used for the administration of all economic transactions; but it is also as a statistical system aiming at measuring profit and other key economic figures. The term *Business intelligence* is often used when corporate data are used for statistical purposes.

Manufacturing enterprises have information systems consisting of two main parts: a financial system and a logistics system for material and production management. These systems contain hundreds of registers with thousands of variables. In the same way as a national statistical office uses administrative registers to create statistical registers, an enterprise can use its information systems to create registers that are used as sources for the enterprise's internal register-based statistics, for example, on sales. These statistical registers contain microdata for all the transactions relating to new orders and invoicing.

Certain registers can be considered as *base registers* in the register system of a manufacturing enterprise. The *items register* and the *client register* are two examples of base registers that define important object types. Important linkage variables in the system would then be the client identity number and item identity number.

As we have described how society's administrative systems can be used for statistical purposes, investigating how an enterprise's register data could be used statistically is also possible. Statistical science should contribute to developments in this field. The need is illustrated by the rapid growth of Business Intelligence and Data Mining.

This is illustrated by an example describing a register survey in a manufacturing enterprise. The survey in the example is the monthly survey on sales.

Every month, a statistical register is created by matchings and selections from three administrative registers (Chart 3.4): the Invoice Register with all transactions regarding invoices, the Client Register and the Item Register.

Chart 3.4   Three administrative registers

Invoice Register

Date	Client number	Item number	Quantity	Value
2012–01–18	196	22	10	832
2012–01–19	28	4	500	20 339
2012–01–19	7	128	40	9 840
2012–01–20	23	9	100	10 622

Client Register

Client number	Seg-ment	Coun-try
7	3	SE
23	3	SE
28	3	SE
196	2	UK

Item Register

Item number	Item group	Pre-calcu-lated cost
4	1	36
9	1	90
22	2	28
128	2	205

A statistical Sales Register (Chart 3.5) is created as follows:

- The Invoice Register is important because it contains a combination of two identities.
- All transactions for one defined month are selected from the Invoice Register to be the statistical units in the Sales Register. This monthly register is matched against the Client and Item Registers, and variables from these registers are imported into the Sales Register.
- The Sales Register for a given month is then used to create tables containing invoiced values at current prices, price indices (item prices compared with item prices during a base year), invoiced volumes (values at constant prices), and gross profit margins by segments, countries and item groups. These tables are used to update a time series database.

Chart 3.5   Statistical Sales Register for January 2012 – four transactions

Date	Client number	Seg-ment	Coun-try	Item number	Item group	Quantity	Value	Price	Pre-calcu-lated cost	Gross profit
2012–01–18	2196	2	UK	122	2	10	832	8.32	280	552
2012–01–19	1028	3	SE	104	1	500	20 339	40.68	18 000	2 339
2012–01–19	1007	3	SE	128	2	40	9 840	246.00	8 200	1 640
2012–01–20	1023	3	SE	333	1	100	10 622	106.22	9 000	1 622

The example in Charts 3.4 and 3.5 shows that the concepts regarding registers and register surveys are of a general nature.

- Base registers with different kinds of administrative object sets can be linked if there are administrative registers with combinations of identities such as the Invoice Register in Chart 3.4.
- Administrative registers can be transformed into statistical registers as illustrated in Chart 3.5.
- The variables in Chart 3.5 do not have the error structure that is common in questionnaire data; for example, measurement errors are not a problem here.

3.2.1   Corrections in accounting data

Administrative registers must be transformed into statistical registers. The administrative data can have good administrative quality, but they may still be unsuitable for statistical purposes. Two examples below illustrate the importance of transforming administrative data. Errors in administrative data should be corrected according to statistical principles, and missing values should be treated as missing values in statistical data.

On 27 February, an invoice transaction is registered in the Invoice Register. When the transactions for February are checked, this transaction is found to be incorrect and should not have been made. A correction is made on 2 March as shown in Chart 3.6.

Chart 3.6   Administrative Invoice Register

Date	Client number	Item number	Quantity	Value
2012–02–27	1053	333	1 000	107 560
2012–02–28	1034	112	655	32 700
2012–03–01	1117	104	500	20 339
2012–03–02	1053	333	–1 000	–107 560

Chart 3.7   Statistical Invoice Register

Date	Client number	Item number	Quantity	Value
2012–02–28	1034	112	655	32 700
2012–03–01	1117	104	500	20 339

According to accounting principles, errors are corrected by adding a new transaction so that the erroneous transaction and the correction will sum up to zero. From a statistical point of view, this corresponds to two errors with different signs. In the statistical register, Chart 3.7, neither the error nor the correction should be included because the time series will have low statistical quality when the error and the correction appear during different months.

3.2.2   Missing values in accounting data

Administrative registers can contain missing values. If these are interpreted as zeros, the statistical analysis will be misleading. When the statistical register is created, missing values should be detected and replaced by imputed values as in Charts 3.8 and 3.9.

Chart 3.8   Administrative Item Register

Item number	Item group	Pre-calculated cost
104	1	36
333	1	
122	2	28
128	2	205

Pre-calculated costs may not have been calculated for earlier months for all items.

If these missing values are not detected, the time series gross margin will be disturbed.

Conclusion: Administrative data must be edited before they are used for statistical purposes.

Chart 3.9   Statistical Item Register

Item number	Item group	Pre-calculated cost
104	1	36
333	1	94.42
122	2	28
128	2	205

Other sources give the following information:

Total sales of items with known pre-calculated costs during the period were SEK 90 000. Pre-calculated costs of these sales were SEK 80 000. Pre-calculated costs are thus 8/9 of total sales

The average selling price of item 333 during the period is SEK 106.22. The imputed pre-calculated cost for item 333 is then 8/9 of average selling price: $106.22 \cdot (8/9) = 94.42$

Register surveys are common within enterprises and other organisations. The methodological problems associated with these enterprise surveys should be investigated and discussed in the same way as we discuss how statistical offices can use administrative data to produce official statistics.

These problems are not often recognised as statistical problems, because statistical science is often only associated with survey sampling, randomised experiments, probability and inference theory.

3.2.3 Administrative and statistical information systems

The use of administrative data for statistical purposes is not specific to national statistical offices. It is a common practice in large enterprises and organisations. Administrative systems are generally used as sources of statistical information. There is no major difference between the following enterprise example and register-based statistics at a national statistical office:

- Statistics on staff and salaries within an enterprise can be produced using the personnel management system.
- Population and income statistics are produced at a statistical office using data from the National Tax Board's tax collection system for population registration and tax assessment.

Certain information systems are built solely for statistical purposes, such as the Labour Force Survey, which is conducted in many countries. Therefore, these systems can be completely designed according to statistical principles.

Other information systems are used for administrative as well as statistical purposes, which can sometimes lead to conflicts with regard to the system's structure. These systems are often primarily intended for administrative purposes and the statistical information is a by-product. However, there are several differences between a pure administrative system and a pure statistical system. These two kinds of systems are compared below.

Different purposes
Information in an administrative system is used as a basis for taking administrative measures and decisions that affect the objects in the system.

Example: A personnel management system is used to make salary payments every month. For each employee, a decision is made regarding how much should be paid for the specific month.

Information in a statistical system is used as the basis for analysis and drawing conclusions regarding groups of units. These conclusions can serve as the basis for policy-related decisions.

Example: A statistical salary system is used to study salary structure. How has this changed? What are the differences in monthly salaries between different staff categories? This analysis could then involve a change in policy relating to salary issues, for example, that women should be better paid.

Different roles for individual objects
In an administrative system, decisions are made and measures are taken with regard to individual objects. Information relating to that specific object is retrieved to this end.

Example: Salaries are paid to every employee in an enterprise. Administrative information is checked and salary and tax for each employee can be calculated.

In a statistical system, the individual objects are not of interest per se. In a statistical analysis, aggregate estimates are calculated and compared for groups of objects.

Example: Salary totals, average salaries, the dispersion of salaries, etc. are calculated for different staff categories.

Approaches regarding errors

From an administrative point of view, certain items of information must be absolutely correct, but other items can be more approximate. From a statistical point of view, errors can exist, but they should be carefully controlled. Attempts are made to reduce the errors, which may significantly affect statistical conclusions. Errors can be accepted in some data, but only if these are considered to have a limited effect.

Example: The personal identification number in a personnel management system must be completely correct from the point of view of salaries and tax administration. Since the Swedish personal identification number contains the date of birth, it can also be used to describe the age structure of the staff. If, for example, 30% of the staff have an incorrect number for the month in their registration number, this would not affect the statistical analysis particularly, although the salary and tax routines would become impossible.

3.3   Measurement errors in questionnaire and register data

Data that have been collected or created by administrative authorities can have a different character. Some data are actually *statistical data*, if the authority wants to produce its own statistics. For example, the Ministry of Health in a Latin American country issues birth certificates to newborns. Many variables are collected regarding the mother and the birth. The mother is interviewed regarding education and occupation, but these variables are uninteresting when taking care of the mother and the baby. These are statistical variables, and because the hospital staff does not use them the quality is not as good compared with medical variables such as weight and height of the baby.

Other kinds of variables are *legally important* – providing the wrong information is illegal and can be punished. Income assessments and tax returns are examples of this kind of data.

A third category of variables represents *decisions made by an authority*. For example, the Tax Board decides on taxable income and the amount of tax that should be paid; a court decides that a person is guilty of violating a certain law and should receive a specific punishment; social authorities decide that a family is entitled to receive some kind of benefit and set the amount of money they will receive.

Among these different kinds of administrative data, statistical data and data of no legal importance usually have the lowest quality. Legally important data and decisions made by an authority are of the highest quality. The quality of administrative data is important for the individual's rights and obligations in contrast to statistical data that do not have any consequences for the respondent.

Administration in a manufacturing enterprise can serve as an example of administrative data that are purely administrative in nature: A customer phones and asks if enterprise X can deliver a certain quantity of a certain commodity. How much will it cost and when can it be delivered? After negotiations, the following administrative data have been created:

Customer identity:	1234
Item number:	87
Quantity:	250
Price:	568 700
Delivery date:	15–04–2021

This kind of administrative data can be used afterwards for a register survey on sales. Note that there is no measurement here and no collection of data – data are generated through decisions during the administrative process. A statistical measurement is quite different – the true values of the variables exist first, and then we measure and collect the data.

3.3.1 Measurement errors

Many important differences exist between error structures found in statistical registers based on administrative systems and those from data collected in sample surveys.

Measurement processes are quite different between these types of surveys. In sample surveys with questionnaires or interviews, the cognitive processes in answering questions are modelled as (Groves et al., 2004, p. 202):

– comprehension of the question;

– retrieval of information;

– judgement and estimation;

– reporting an answer.

Are the same cognitive processes relevant when persons report data to an administrative authority, either as a private person or representative of an enterprise? These cognitive or psychological processes also exist in connection with administrative reporting, but we believe they are unimportant. Instead, administrative rules and legislation are important factors; and when reporting data from enterprises, accounting principles and practice are more important than psychology. Chart 3.10 compares the two ways of collecting data:

Chart 3.10   Measurement errors – comparison of data collection methods

Collecting data in sample surveys	Collecting data in administrative systems
Underlying structure of question:	Underlying structure of question:
Will you please try to understand our questions and try to remember?	*1. Report last month's turnover before the 12th of this month.*
It is not necessary for you to answer, and it does not matter what you answer, as there is no penalty involved.	*2. Pay 25% of reported turnover before the 12th of this month.*
	3. If you do not report and pay, you will be liable to pay a penalty.
Questionnaire to persons:	Reports to authorities from persons: In our yearly tax form, we only add our signature and perhaps apply for some deductions.
Does the right person in the household answer?	
Questionnaire to enterprises:	Reports to authorities from enterprises:
Does the right person within the enterprise answer?	Regular duty of professionals, the enterprise's accounting system as a rule generates the report.
	Errors are errors in the accounting routines or typing or scanning errors.
Interviewer effects can be disturbing	No interviewers, no interviewer effects
Leading questions in market research are often a problem.	Legally complicated questions
Variables collected via questionnaire or interview in sample surveys or censuses: measurement errors are important	Statistical variables in register surveys are often derived variables based on administrative variables: relevance errors and model errors are important.

Note that some questions or variables collected in an administrative system are legally important, while other questions are less important. The quality of these unimportant questions can be lower – the respondents can answer whatever they want. There are no consequences and the preconditions are the same as for a question in a sample survey.

3.3.2   Taxation errors
Of course, errors also exist in tax data. The Swedish National Tax Agency analyses tax data and tries to estimate the difference between what the taxpayers actually report and pay and what they should have reported and paid. Statistics Sweden's department for National Accounts also carries out top-down estimates.

Example: A common opinion is that taxpayers only submit data that serve their purposes and consequently tax data are of low quality. In the example below, taken from a leading Swedish newspaper, most people would like to pay as little tax as possible, so the deductions may be higher than are justifiable.

---

80 per cent of Swedish people's tax deductions are pure tax evasion

Taxpayers submit errors worth billions in their tax declarations. Complicated rules and unclear legislation have made it hard for the country's tax authorities to check all the deductions. Errors can be found primarily in the deductions for share transactions, management fees and other share-related charges.

Deductions for the sales of shares

– 1/3 of all share sales contain errors

Deductions for management fees

– 125 000 taxpayers claimed deductions of a total of SEK 515 million
– 66% of these deductions contain errors. Tax errors can be calculated at SEK 90 million

Deductions for other expenditure

– 700 000 taxpayers claimed deductions of a total of SEK 2 800 million
– 82% of these deductions contain incorrect information
– Tax errors can in total be calculated to amount to around SEK 700 million

---

The headline exaggerates in several ways, '80 per cent' is an exaggeration and 'tax evasion' is often based on misunderstanding due to complicated rules:

- Deductions for share sales: the errors are largely unintentional.
- Deductions for management fees: the errors are on average 17% (90/515) and the most common error may be unintentional due to misunderstanding.
- Deductions for other expenditure: 82% of these deductions contain errors but the deductions are on average 25% (700/2 800) incorrect.

Another perspective on these errors is when they are compared to the total income for all those filing tax declarations, the error is 0.3%.

The fact that deductions in the declarations are too high, and that consequently the tax is too low, does not mean that the statistics in the Income Register are of low quality, even though they are based on these declarations. Assume that we have data for a person who makes excessively high deductions on her/his tax declaration, but otherwise declares correctly:

Income from employment	257 600	The income is correct.
Deductions for other expenditure	25 500	The deduction is too high but is accepted.
Tax	100 000	The tax is too low according to the rules, but statistically correct, as this is the tax that the person actually paid.
Disposable income	157 600	Statistically correct

Statistics Sweden's statistics on disposable income are not incorrect because of this person's data. To collect income data with questionnaires or interviewers is difficult and measurement errors would be a more serious problem than in tax data.

CHAPTER 4

Building the System – Record Linkage

When the transition from a census-based system into a register-based starts in a country, the following areas will become relevant first:
- How to handle record linkage or micro integration of registers?
- How to assess the input quality of the administrative registers and the output quality of the new statistical registers?
- How to create the base registers for social and economic statistics?

Chapters 4–7 and 10 treat these topics.

4.1 Record linkage

This is a commonly accepted term for finding observations in two database tables, where the statistical units have the same identities or *matching keys*. Matching, record linkage and micro integration are related terms for the same concept. We distinguish between two types of matching:

1. The purpose of matching is to find *identical objects* in different registers or database tables. When matching, one or a combination of several identifying variables in the relevant registers are used as *matching keys*.

   Example: Two registers on individuals are matched; the matching key PIN, personal identification number, exists in both.

   Example: Two registers on individuals are matched with a combination of first name, surname and birth date.

2. The purpose of matching is to find objects that have a *certain type of relation* to one another. These objects can be found within the same database table or in different registers concerning different object types. When matching, a *reference variable* from the first register and a corresponding variable from the second register are used as the matching key.

   Example: A register on individuals contains the identifying variable personal identification number as well as a reference variable giving the personal identification number of that person's spouse. Two copies of this register are

Register-based Statistics: Registers and the National Statistical System, Third Edition. Anders Wallgren and Britt Wallgren.

matched using the personal identification number as the primary key in the first copy and the reference variable as a foreign key in the other copy.

Why is record linkage so important?

There are many reasons why record linkage is necessary for register-based statistics:

- In new register countries, combining many registers is often necessary to obtain *good coverage* of the population.
- A common practice in the Nordic countries involves combining registers to obtain a *rich variable content.*
- When many registers have been combined, there can be many *duplicates* that must be found and removed
- Many *quality issues* are revealed when registers are combined or when sample surveys and registers are combined at the micro level. Coverage can be compared and similar variables in different sources can also be compared. Through these comparisons, errors and quality issues will be found and errors can thereafter be reduced.
- When registers are combined, new variables can be created by *adjoining* and *aggregation.*

4.2   Record linkage in the Nordic countries

Record linkage in the Nordic countries is carried out by matching observations in different sources. Only one identifying variable is used when two sources are combined – a national identity number or a code for a standardised address. Because these identities precisely define one statistical unit, the linkage method based on matching keys is practically without errors.

The Nordic NSOs use this method of record linkage daily in the production of official statistics. This method is also used to combine about 150 population registers in Statistics Sweden's largest longitudinal register that is used by researchers. The linkage method is a very efficient survey method because the quality of these large-scale record linkage operations is high. The method is seemingly simple – 'only' deterministic record linkage with one matching key without errors is used and all links are one-to-one. However, developing a national statistical system with this capability represents a very advanced example of survey system design.

We describe this matching method as the *register system approach.* The work is organised in a specific way to deal with quality issues:

- Special units at Statistics Sweden are responsible for each *base register.* A base register is used to create one or more *standardised populations* that are used by other units that work with register-based statistics.
- The unit responsible for the Population Register, for example, is also responsible for the PIN variable. This means that the unit staff must keep track of persons who are allowed to change their PIN or replace a prelimi-

nary PIN with a definitive PIN. Old and new PINs are included in a cross-reference table together with the date when the change occurred. Given this information, the staff at the Population Register can edit administrative data on persons by replacing the PINs in the source with corrected PINs.

– A small number of persons with incorrect or unknown PINs are excluded when the standardised populations are created that are used by other register units. These persons may be foreigners studying at Swedish universities, and they will be excluded from the population when only persons who permanently live in Sweden are included.[1]

Comparisons with other countries

The probabilistic record linkage methods developed by Fellegi and Sunter (1969) are not used in the Nordic countries. Their probabilistic matching methods were developed for use in situations where there are no high-quality identity numbers that are used in many sources. Such situations require combining a number of variables such as name, address, birth date and birthplace.

If we compare record linkage in the Nordic countries and the US, for example, we find that quite different methods are used today. The probabilistic linkage methods used in the US require specialised competence and difficult work, and they are used with the intention of matching a moderate number of files. Thousands of registers are matched every year in the Nordic countries, which would not be possible with probabilistic methods.

Quality of the matching results is another aspect. A statistician from outside the Nordic countries may be happy if 90% of the observations are correctly matched. However, if you want to create a longitudinal register with data on persons and must match more than 150 registers, then 10% errors in each matching will result in a register that is worthless.

New register countries – long-term and short-term strategies

The long-term strategy should be to develop the national administrative systems so that national identity numbers are used in all administrative registers. *One* authority should be responsible for the central population register used by all other authorities. The national statistical office should then have access to these administrative data under the precondition that integrity is protected. With these preconditions, the statistical system can gradually develop so that record linkage can be carried out using the Nordic register system approach.

National identity numbers for enterprises are already available in many countries, and deterministic record linkage can be used for economic statistics.

The short-term strategy may require combining a limited number of registers with the methods for probabilistic record linkage described in Winkler (1995, 2006, 2008) and Herzog, Scheuren and Winkler (2007) and Dasylva et al. (2019). These methods are briefly discussed below.

---

[1] We recommend the modification of this practice to allow more flexible methods of defining populations, for example, by including workers from other EU countries that otherwise would be excluded from the Swedish LFS.

4.3 Deterministic record linkage

Deterministic record linkage compares the observations in two different registers. Both registers contain the same *matching key*, consisting of one or a combination of identifying variables. All observations in each register are identified by the matching key, and there are no duplicates with the same values on the matching key in the same register. With the matching key, pairs of *observations* in the two registers are *linked* if the two observations have *exactly* the same values on the matching key. We also say that the *registers* are *linked* by this matching key.

Chart 4.1a   Matching without errors in the matching key (PIN)

A	B	New register		
PIN	PIN	from A	from B	
11111	11111	11111	11111	match
22222	22222	22222	22222	match
33333	44444	33333	null	non-match
44444	55555	44444	44444	match
55555	66666	55555	55555	match
77777	77777	null	66666	non-match
88888	88888	77777	77777	match
99999	99999	88888	88888	match
		99999	99999	match

Chart 4.1a uses PIN as the matching key. As PIN is the primary key in both registers, we are conducting a one-to-one match. Although there are no errors in the matching key, we obtain a non-match for two observations. This will often occur when registers are matched. Due to *differences in coverage*, there will be a number of non-matching observations even if the matching key is of good quality. When record linkage is conducted at the Nordic NSOs, we know that obtaining observations that do not match is due to differences in coverage.

Chart 4.1b   Matching with errors in the matching key (PIN)

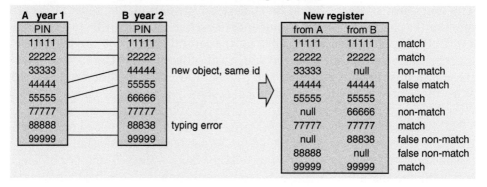

A year 1	B year 2		New register		
PIN	PIN		from A	from B	
11111	11111		11111	11111	match
22222	22222		22222	22222	match
33333	44444	new object, same id	33333	null	non-match
44444	55555		44444	44444	false match
55555	66666		55555	55555	match
77777	77777		null	66666	non-match
88888	88838	typing error	77777	77777	match
99999	99999		null	88838	false non-match
			88888	null	false non-match
			99999	99999	match

Chart 4.1b shows errors in the matching key. This is the only difference compared with Chart 4.1a. Due to these errors, we now obtain *false matches* and *false non-matches*.

Note that when we match two registers with identity numbers of good quality, we can interpret the non-match as differences regarding *coverage* as in

Chart 4.1a. If two registers are matched with a matching key of medium quality, the non-matching observations are often interpreted as *matching problems* generated by errors in the matching key. However, they may have been generated by a combination of coverage errors and matching problems generated by observations with errors in the matching key.

The proportion of false matches and false non-matches can be estimated by taking a sample of matching observations and another sample of non-matching observations. Then the samples are analysed using all information to decide if matches respective non-matches are true or not. Thereafter, matching errors and coverage errors can be described.

In Chart 4.1c, the matching key consists of two variables: each person's first name and surname. When we use text strings such as names or addresses as matching keys, problems with spelling errors and spelling variations may occur that generate false matches and false non-matches. The name *Dd Ddd* seems to be a popular name as two different persons have this name. This generates a false match. The person *Gg Ggg* is spelled as *G. Ggg* in register A, and this generates two false non-matches. Before matching, spelling errors should be corrected if possible, and address formats should be standardised.

Chart 4.1c   Matching with errors in the matching key (First name *and* Surname)

A			B			New register				
first name	surname		first name	surname		from A		from B		
Aa	Aaa		Aa	Aaa		Aa	Aaa	Aa	Aaa	match
Bb	Aaa		Bb	Aaa		Bb	Aaa	Bb	Aaa	match
Cc	Ccc		Dd	Ddd	not same	Cc	Ccc	null	null	non-match
Dd	Ddd		Ee	Eee		Dd	Ddd	Dd	Ddd	false match
Ff	Fff		Ff	Fff		null	null	Ee	Eee	non-match
G.	Ggg		Gg	Ggg	typing error	Ff	Fff	Ff	Fff	match
Hh	Hhh		Hh	Hhh		G.	Ggg	null	null	false non-match
						null	null	Gg	Ggg	false non-match
						Hh	Hhh	Hh	Hhh	match

In the above examples, we used identities regarding the *same* persons in the registers we want to match. In Chart 4.1d, the *PIN of persons* is the primary key and *PIN of the person's spouse* is a foreign key. As each person can have at most one spouse, this will generate one-to-one links.

Chart 4.1d   Matching without errors in the matching key (PIN)
PIN of spouse (foreign key) in register A *and* PIN (primary key) in register B

A			B			New register			
PIN	PIN spouse		PIN	Age		PIN	PIN spouse	Age	Age spouse
11111	null		11111	21		11111	null	21	null
22222	null		22222	26		22222	null	26	null
33333	88888		33333	35		33333	88888	35	37
44444	null		44444	54		44444	null	54	null
55555	null		55555	31		55555	null	31	null
66666	99999		66666	41		66666	99999	41	44
77777	null		77777	67		77777	null	67	null
88888	33333		88888	37		88888	33333	37	35
99999	66666		99999	44		99999	66666	44	41

One-to-many links will be generated if we use the PIN of each person's mother as the foreign key, since the same mother can have more than one child. Note that in Chart 4.1d we are matching different object types – *persons' spouses* with *persons*. As a result, a new derived variable (Age spouse) was created.

4.4   Creating variables by adjoining and aggregation

Creating adjoined or aggregated variables involves matching registers containing different object types that have a certain relation to one another. Variables for one kind of objects are transformed via this relationship into variables for another kind of objects. When the relation is *one-to-one*, this transformation is simple; when the relation is *one-to-many* or *many-to-one* the transformation is more complex and consists of adjoining or aggregation.

Variables derived by adjoining

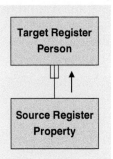

This involves creating a derived variable in a register using variables from another register. The objects in the source register can be linked to objects in the target register in a *one-to-one* relation or a *one-to-many* relation. This means that every object in the source register can be linked to one or many objects in the target register. Using this relationship, variables in the source register can be *adjoined* to the objects in the target register.

Example: In a register on individuals, the geographical coordinates of the property (or dwelling) can be adjoined to each person. Registers on individuals should contain the identity of the property where the person is registered. The property identity is the link to the Real Estate Register. The property's coordinates are transferred from the Real Estate Register to the relevant register on individuals. Here, properties and individuals are linked in a *one-to-many* relation, where one property (or dwelling) is linked to one or many individuals.

Variables derived by aggregation

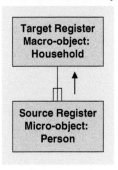

This also involves creating a derived variable in a register using variables from another register. The objects in the source register can be linked to the objects in the target register using a *many-to-one* relation. One or many objects in the source register can be linked to one object in the target register. Values can be *aggregated* in a relevant way for the survey for the *micro-objects* in the source register that are linked to the specified *macro-object* in the target register.

Example: Household income is an aggregated variable formed by adding the values of the variable income of individuals for all individuals in a certain household. Household is the macro-object and person is the micro-object.

Example: How can information from a register on employees be combined with information from a register on enterprises? For enterprises, a derived variable is formed, *share of employees with higher education*. This variable for the macro-object enterprise is formed by calculating the share of persons with higher education among all the employees (the micro-objects) at the enterprise.

Integrating registers with different object types

Adjoining and aggregation are explained by the following example. We start with three registers in Chart 4.2a before any matching or creation of derived variables has been done. One person can have many jobs and one establishment can have many employees. *Wage sum* is used as the name for three different variables:

- Wage sum for *job*, gross annual pay that one person has at one job.
- Wage sum for *person*, aggregated gross annual pay for all jobs of one person.
- Wage sum for *establishment*, aggregate gross annual pay for all jobs at one establishment.

The example illustrates how data from three registers with different kinds of object types are integrated. Seven new variables are created by adjoining and aggregation.

Chart 4.2a   The relations between persons, activities and establishments

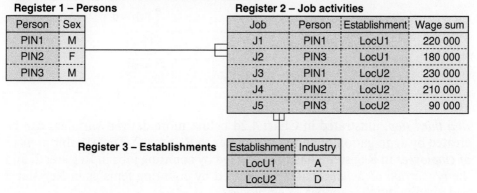

Register 1 – Persons

Person	Sex
PIN1	M
PIN2	F
PIN3	M

Register 2 – Job activities

Job	Person	Establishment	Wage sum
J1	PIN1	LocU1	220 000
J2	PIN3	LocU1	180 000
J3	PIN1	LocU2	230 000
J4	PIN2	LocU2	210 000
J5	PIN3	LocU2	90 000

Register 3 – Establishments

Establishment	Industry
LocU1	A
LocU2	D

In the first step illustrated in Chart 4.2b, wage sums for persons and establishments are derived by aggregation of wage sums for job activities. Data for jobs are aggregated into one value for each person or establishment.

In Chart 4.2b there are three different *wage sum* variables defined for three different object types – persons, job activities and establishments.

Chart 4.2b   Wage sums for persons and establishments created by aggregation

Register 1 – Persons

Person	Sex	Wage sum
PIN1	M	450 000
PIN2	F	210 000
PIN3	M	270 000

Aggregation

Register 2 – Job activities

Job	Person	Establishment	Wage sum
J1	PIN1	LocU1	220 000
J2	PIN3	LocU1	180 000
J3	PIN1	LocU2	230 000
J4	PIN2	LocU2	210 000
J5	PIN3	LocU2	90 000

Aggregation

Register 3 – Establishments

Establishment	Industry	Wage sum
LocU1	A	400 000
LocU2	D	530 000

In a second step, derived variables for the job activities in Register 2 can be created by adjoining variable values from Registers 1 and 3. This is illustrated in Chart 4.2c below.

Chart 4.2c   Industry and sex as derived variables for jobs created by adjoining

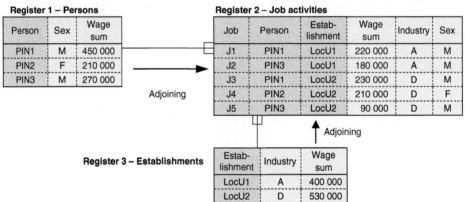

Register 1 – Persons

Person	Sex	Wage sum
PIN1	M	450 000
PIN2	F	210 000
PIN3	M	270 000

Adjoining

Register 2 – Job activities

Job	Person	Establishment	Wage sum	Industry	Sex
J1	PIN1	LocU1	220 000	A	M
J2	PIN3	LocU1	180 000	A	M
J3	PIN1	LocU2	230 000	D	M
J4	PIN2	LocU2	210 000	D	F
J5	PIN3	LocU2	90 000	D	M

Adjoining

Register 3 – Establishments

Establishment	Industry	Wage sum
LocU1	A	400 000
LocU2	D	530 000

In a third step, illustrated in Chart 4.2d below, more derived variables can be created by aggregation of Industry and Sex in Register 2. The variable *number of employees* in Register 3 has been created by counting jobs in Register 2, and the *proportion of females* has been created by counting females in Register 2 and dividing by the number of employees.

Up to now, all aggregations have been straightforward. However, when *Industry* for job activities is aggregated into Industry for persons, the aggregation of the qualitative variable Industry is more complicated. For each person in Register 1, the *type value* of Industry in Register 2 is computed with the wage sums in Register 2 as weights or frequencies.

Chart 4.2d　Industry, number of employees and proportion of females as derived variables – by aggregation

Register 1 – Persons

Person	Sex	Wage sum	1st Industry
PIN1	M	450 000	D
PIN2	F	210 000	D
PIN3	M	270 000	A

Aggregation

Register 2 – Job activities

Job	Person	Establish-ment	Wage sum	Industry	Sex
J1	PIN1	LocU1	220 000	A	M
J2	PIN3	LocU1	180 000	A	M
J3	PIN1	LocU2	230 000	D	M
J4	PIN2	LocU2	210 000	D	F
J5	PIN3	LocU2	90 000	D	M

Aggregation

Register 3 – Establishments

Establish-ment	Industry	Wage sum	No. empl	Prop F
LocU1	A	400 000	2	0.00
LocU2	D	530 000	3	0.33

In the third step, inconsistencies were created in this system of three registers. The total number of employees is *three* in Register 1, but *five* in Register 3. Wage sums by Industry in Register 1 differ from wage sums by Industry in the other registers. The conclusion is that matching and creating derived variables can raise difficult methodological issues, which are discussed later in the book.

4.5　Probabilistic record linkage

With deterministic matching, two observations are classified as linked if the values on the matching key are *exactly* the same. This principle of exact agreement is not suitable when the matching key consists of names, addresses, birth dates, etc. If exact agreement is required for a link between observations, small insignificant differences in the matching key will result in a high proportion of false non-matches. There are a number of methods that can be used to reduce such matching errors. Short descriptions of these methods are presented here.

Birth data with mothers of twins
We illustrate some of the methods with data on births during one year in a Latin-American region. The Birth Register consists of 35 252 births, of which 527 are twins (or triplets).

Data on a number of potential matching variables were collected by the hospital staff together with medical data. We use these potential identifiers to investigate if we can find the duplicates or twin mothers in the register. Hospital staff filled in a paper form by hand for each birth. This was later edited by the Ministry of Health.

The data were collected with the intention of producing health statistics, and this is possible without combining these data with other sources. This is actually an example of administrative data that are similar to statistical data as discussed in Section 3.3. But if the identifiers are sufficiently good, these health data could be combined with other sources to produce other kinds of statistics.

The data for the two observations for each twin mother seem to have been collected independently, as we find a number of differences between the observation for the first and second twin regarding variables describing the same mother. One variable contains information on twin births which we have used to identify the twin mothers, but this variable will not be used in the matching exercise. We discuss the following potential matching variables:

The mother's first name, second name, first surname and second surname

Date of birth of the mother; region and municipality where the mother was born

Residential address, region, municipality, and district; mother's civil status and level of education

Standardisation

Chart 4.3 contains the names and addresses related to 14 twin mothers. A data collection method that generates dirty data should be improved. Chart 4.3 clearly shows that the data collection method generates problems for statisticians who want to link observations.

All identifiers should be edited, and spelling errors should be reduced. In Chart 4.3 'MARYLIN' is probably misspelled. The nurse asks for the name and writes what she has heard. An alternative could be to let the mother write her own name, which could reduce misspellings.

Addresses should be standardised as much as possible. Only one question appears in the form used for the data in Chart 4.3:

Residential address: ....

A better method would be to ask four questions:

Residential address:

Street name: ...;   Street number: ...;   Postal code: ...;   City: ...;

The pattern of names and addresses depends on language and country; this means that nation-specific methods should be developed. Chart 4.3 consists of names and addresses related to 14 twin mothers numbered from 1 to 14. Names 1–4 have the same pronunciation and very small differences in spelling. Names 5–7 differ in M and N and names 8–10 differ more.

There are four addresses; street name is written in five ways (C.14,..., C- 43) and street number in eight ways (X 25, ..., N. 188).

Chart 4.3   Spelling errors, spelling variations and unstandardised addresses

First name	Surname 1	Surname 2	Residential street address
1: MARILYN	3: POL	7: CAMAL	11: C.14 X 25 #107
MARYLIN	POOL	CANUL	C.14 X25 #107
2: MARIA ESTER	4: BAAK	8: CHE	12: CALLE 26 NO. 240
MARIA ESTHER	BAK	CHUC	CALLE 26 NO.0240
	5: BATUM	9: CANCHE	13: 26 NJUM 67
	BATUN	CHE	26 NUM 67 COLONIA TAMARO
	6: COCOM	10: MENDE	14: C-43 No. 188 X 24 Y 28
	COCON	MENDEZ	C- 43 N. 188 X 24 Y 28

The chart illustrates that measuring differences between names is important so that small differences have a small impact on the likelihood of being classified as a link. Strings can be compared, and scales have been developed that measure phonetic similarity. Matching errors can be reduced if these scales are used instead of the crude measure *exactly same/not exactly same*. In addition, lists with aliases can be used where nicknames and names are linked, and synonyms for occupation, for example, can be linked in such lists. Teacher, professor, lecturer, instructor can be linked as similar.

The chart also shows that addresses must be *standardised*, otherwise address information cannot be used for matching purposes. About 50% of the twin mothers have typographically different street addresses in the observations regarding the first twin and the second twin.

Using more variables

Choosing variables and deciding on the number of matching variables to be used are important decisions when designing the matching method. Better variables will reduce matching errors. More variables used for matching can reduce false matches, but they increase false non-matches. Different sets of variables may be more or less suitable depending on whether the registers that should be matched are close in time or not. Place of residence can be a good matching key if the registers are close in time, but a bad variable if they are not.

The three names, *First name*, *First surname* and *Second surname* should be used, as they have good discriminating power. Since we have regional registers in the case with twin mothers, the variables region and municipality of birth, and residential region, municipality and district are less suitable. Almost 50% of the mothers are born and live in the same district.

Civil status has low discriminating power as about two-thirds of the mothers are married. Level of education, however, could be used to distinguish between mothers, as that variable has ten distinct levels. Because the two registers were created during the same time period, variables such as level of education and residence can be useful as matching variables.

Parsing

One difficult variable can be transformed into a number of simple variables by dividing strings containing addresses, for example:

Address = Street name + Street number + Postal code + City

One string variable is divided into four variables that can be treated separately, and the effect of small typing errors and variations can be reduced.

A number of variables could be parsed in the example with twins. Date of birth could be parsed into three variables: day, month, and year. Hence, a difference in one of these will only affect one variable that will indicate non-match; the two others can indicate match. Place of birth and residence should also be parsed into region and municipality of birth, and residential region, municipality and district. Then the information in these variables is utilised better, and the rates of false match and false non-match can be reduced.

Blocking

Record linkage may require heavy data processing. If N observations from register A are matched with M observations in register B, then $N \cdot M$ comparisons are required where N and M as a rule are millions of observations. Blocking can be used to reduce the processing time. One or a few variables can be used to divide each register into a number of smaller registers for each category of the blocking variables. However, if these blocking variables do not have perfect quality the number of false non-matches will increase. Another reason for dividing the registers into smaller, more homogeneous, registers is that different variables are available for different parts of the population.

One variable could be used for blocking in the example with the twin mothers. About 50% of the mothers live in the same district. If this information is used to create two blocks, then the matching operations will be halved. District is a variable of good quality; only one twin mother will be classified as non-match because the mother of the twins lives in different districts according to the recorded data. If we use level of education as a blocking variable, the matching operations would be much reduced; but as coding is uncertain, we will obtain more twin mothers that become false non-matches.

Searching for duplicates

This exercise searches for duplicates in the Birth Register, where the mothers with twins or triplets are duplicates. It is easy to find observations that have the same values with identifying variables as names and birthdates. Depending on which variables we use for identification and depending on errors in these variables, we obtain false duplicates and missing duplicates. Chart 4.4 shows the results for two ways of choosing identifying variables.

Chart 4.4 Two ways of searching for duplicates in a Birth Register

	Four identifying variables: 1st name, 1st surname, 2nd surname, birthdate			**Five** identifying variables: 1st name, **2nd name**, 1st surname, 2nd surname, birthdate		
Mothers with:	One obs.	Two obs.	Three obs.	One obs.	Two obs.	Three obs.
Single baby	**34 658**	65	2	**34 666**	59	0
Twins	158	**353**	4	184	**331**	0
Triplets	4	2	**6**	4	2	**6**

Single-baby mothers should be represented by one observation, twin mothers by two observations, and mothers who have triplets by three observations. The bold numbers in the diagonals are correct. The numbers above the diagonal are false duplicates and the numbers under the diagonals represent missing duplicates. When one more identifying variable is added (2nd name) the number of false duplicates is reduced, and the number of missing duplicates is increased.

The quality of the matching keys is not sufficient in this case; and finding and eliminating the duplicates in the Birth Register are therefore difficult.

Likelihood ratio-based link decision rules – choosing matching keys
Probabilistic record linkage is based on comparisons of probabilities of whether or not pairs of observations match. The following exercise uses data from the same birth register that was used in the exercise of finding duplicates. The birth register has now been cleansed – about 200 observations have been omitted.

Assume that we want to match two registers A and B with birth data using a matching key consisting of three names *1st name, 1st surname* and *2nd surname*.

Register A: 17 539 observations: 176 twin mothers (1st twin) and 17 363 single-baby mothers

Register B: 17 538 observations: 176 twin mothers (2nd twin) and 17 362 single-baby mothers

We measure if the three names are the same for all combinations of units from A and B. Different probabilities are calculated in Chart 4.5. In all, 307 598 982 pairs are compared. The matching set *M* consists of 176 pairs with matching twin mothers, and the non-matching set *U* consists of 307 598 806 pairs. We used the two registers and calculated the probabilities in Chart 4.5.

We found one twin mother where the first name was spelled in different ways in the observations for her two twins. As we have a total of 176 twin mothers, we have that P (*1st name* agrees for a randomly chosen matching pair) = 175/176. The *1st surname* agreed for 166 twin mothers and the *2nd surname* agreed for 168 twin mothers.

We matched the 17 539 – 176 single-baby mothers in A with the 17 538 – 176 single baby mothers in register B.[2] When we matched with *1st name* we obtained 170 621 matches; when we matched with *1st surname* we obtained 1 782 572 matches; and with *2nd surname* we found 1 870 235 matches. By dividing by *U* = 307 598 806, we obtain the probabilities that the *1st name*, etc. agree for a randomly chosen non-matching pair. We have calculated all probabilities in Chart 4.5 in this way.

Chart 4.5 Calculation of probabilities for outcomes for *1st* name, *1st* and *2nd* surname

Register A:	17 539 observations	17 539 · 17 538 = 307 598 982 pairs are compared:
Register B:	17 538 observations	176 pairs are matching (*M*)
		307 598 806 pairs are not matching (*U*)

M: matching pairs	*1st* name	*1st* surname	*2nd* surname
agree Yes	0.994318 = 175/176	0.943182 = 166/176	0.954545 = 168/176
agree No	0.005682 = 1 – P(Yes)	0.056818 = 1 – P(Yes)	0.045455 = 1 – P(Yes)

U: non-matching pairs	*1st* name	*1st* surname	*2nd* surname
agree Yes	0.000555 = 170 621/*U*	0.005795 = 1 782 572/*U*	0.006080 = 1 870 235/*U*
agree No	0.999445 = 1 – P(Yes)	0.994205 = 1 – P(Yes)	0.993920 = 1 – P(Yes)

---

[2] This is not the whole non-matching set *U*. Twin mothers in A should be matched with single mothers in B, etc. Of 307 598 806 comparisons in *U*, we use only 17 363 · 17 362 = 301 456 406 comparisons; but we obtain good approximations of the matching probabilities for the set *U*.

For each pair of observations, we obtain the eight possible outcomes in Chart 4.6. Assuming that the outcomes are independent for the matching set as well as the non-matching set is customary.

P (all three names agree for a randomly chosen matching pair) =

= 0.994318 · 0.943182 · 0.954545 = 0.895195

The probabilities in column (2) are calculated in a similar way.

P (all three names agree for a randomly chosen non-matching pair) =

= 0.000555 · 0.005795 · 0.006080 = 0.00000002

Chart 4.6   The eight outcomes for the matching key: 1st name, 1st and 2nd surname

	Comparison of one unit from A and one from B			Probability(outcome)		Likelihood
	1st name agree	1st surname agrees	2nd surname agrees	if match (1)	if non-match (2)	ratio (R)
1.	Yes	Yes	Yes	**0.895195**	**0.00000002**	45 803 177
2.	Yes	Yes	No	0.042628	0.000003	13 342
3.	Yes	No	Yes	0.053927	0.000003	16 083
4.	No	Yes	Yes	0.005115	0.000035	145
5.	Yes	No	No	0.002568	0.000548	4.69
6.	No	Yes	No	0.000244	0.005757	0.042
7.	No	No	Yes	0.000308	0.006042	0.051
8.	No	No	No	0.000015	0.987612	0.000015

If we divide column (1) by column (2) in Chart 4.6, we find the likelihood ratios that are used for determining links in probabilistic matching. The likelihood ratios R are used in a decision rule for each pair of units:

– If $R >$ Upper limit, then the pair is considered a link.
– If Lower limit $\leq R \leq$ Upper limit, the pair is considered a possible link.
– If $R <$ Lower limit, then the pair is considered a non-link.

The upper and lower limits are set to give desired risks for false matches and false non-matches. The pairs with ratios between the limits can be checked manually. The error rates can be estimated from samples of pairs, mainly from possible links.

In the case with twins, we use $R > 20\ 000$ as Upper limit and $R < 20\ 000$ as Lower limit. This matching rule gives 25 false matches, 158 true matches and 18 false non-matches.

If we use $R > 10\ 000$ as Upper limit and $R < 10\ 000$ as Lower limit, we obtain 2 416 false matches, 174 true matches and 2 false non-matches.

The conclusion is that we should try to improve the matching of these registers and also use dates of birth (day, month and year) as three matching variables together with the three names.

After calculating probabilities in a similar manner as in Chart 4.5, the following likelihood ratios are obtained for different outcomes to be used in the rule for classifying pairs as linked or not linked.

Chart 4.7   Some outcomes for the matching key: Name and date of birth of the mother

Comparison of one unit from A and one from B	Probability(outcome)		Likelihood
Variables that agree within a pair of observations	if match (1)	if non-match (2)	ratio (R)
1. 1st name, 1st surname, 2nd surname, day, month, year agree	**0.820679**	0.0000000000	322 728 548 402
1. 1st name, 1st surname, 2nd surname, day, month, agree	0.019086	0.0000000001	368 824 708
1. 1st name, 1st surname, 2nd surname, day, year agree	0.009433	0.0000000000	337 229 413
1. 1st name, 1st surname, 2nd surname, month, year agree	0.044228	0.0000000001	599 743 328
2. 1st name, 1st surname, day, month, year agree	0.039080	0.0000000004	94 010 907
2. 1st name, 2nd surname, day, month, year agree	0.049439	0.0000000004	113 322 420
2. 1st surname, 2nd surname, day, month, year agree	0.004690	**0.0000000046**	1 023 501

Chart 4.7 lists some of the outcomes that could be indicators of a link between two observations. We decide to use the four cases marked 1 as a decision rule for a match. This gives the Upper limit for $R > 300\ 000\ 000$. This matching rule gives 7 false matches, 157 true matches and 19 false non-matches.

If we use $R > 1\ 000\ 000$ as Upper limit and $R < 1\ 000\ 000$ as Lower limit, we will also use the cases marked 2 in Chart 4.7 as a decision rule for a match. Then we have 29 false matches, 171 true matches and 5 false non-matches.

We conclude that using the dates of birth as three matching variables together with the three names considerably improved the quality of the matching.

If the pairs with $1\ 000\ 000 \le R \le 300\ 000\ 000$ are inspected, judging whether there is a match or non-match will be possible. Then all variables can be compared, and typing errors in residential addresses will not be a cause of concern – the human mind can see similarities that the computer cannot find.

In practice, the true links are not known. Thus, the probabilities must be estimated. In the example with birth data, estimating probabilities for non-matching pairs was easy. Error rates among matching pairs were estimated with the true matching pairs. In a real application, these probabilities must be estimated with special estimation methods first developed by Fellegi and Sunter (1969). These methods are described in Herzog, Scheuren and Winkler (2007). Special software for probabilistic record linkage mentioned in Herzog et al. must be used.

4.6   Four causes of matching errors

Using the matching key, which can consist of one or a collection of identifiers, we obtain links between observations in different registers. We use these links to create the new observations in the new combined register. There are four kinds of matching errors that can make the new combined observations erroneous.

- Matching key has errors
- Matching key has changed
- Statistical units are wrong
- Statistical units have changed

These four matching errors are discussed below, and different kinds of errors are illustrated with case studies. Probabilistic matching methods have been developed to deal with the first cause of matching errors. Our contribution here is to discuss the three other causes of matching errors that are problems in all countries, including countries where national identity numbers have good quality.

Matching key has errors

The probabilistic methods discussed above can be used to base the linkage on the efficient use of more variables. With probabilistic record linkage, there can be errors in the matching keys.

Matching key has changed

Countries with good national identity numbers should organise the work in such a way that the base registers are responsible for the identity numbers. The base registers should receive information on changed identity numbers from the administrative authority in question so that old and new numbers can be linked.

Countries without good national identity numbers should handle this problem through record linkage based on many variables. If addresses have changed, there are other variables that can be used to link observations for persons that have moved. Birth date and birth region will be important for the quality of the matching of registers regarding different years. With probabilistic record linkage with many variables in the matching keys, some of the variables can have changed but record linkage is still possible.

Statistical units are wrong

When data regarding different object types but with the same identity numbers are combined, we obtain another kind of matching error. In Section 3.1.2 there is one example where data from the three sources listed below are compared. The sources use the same BIN identity numbers and the three registers have been matched.

BIN	=	Business identity number of each legal unit
SBS	=	Turnover according to Statistics Sweden's questionnaire
YIT	=	Turnover according to the yearly income tax returns
VAT	=	Turnover according to 12 monthly VAT returns

Chart 4.8   Yearly turnover for the same legal units in three sources, USD million

4.8a. Incomplete composite enterprise units

BIN	SBS	YIT	VAT
160006	2 301	0	0
160007	2 211	0	2 239
160008	1 316	1 316	0

4.8b. Complete composite enterprise units (EU)

EU id	BIN	SBS	YIT	VAT
EU1	160006	2 301	0	0
EU1	160666		2 352	2 297
EU2	160007	2 211	0	2 239
EU2	160777		2 203	0
EU3	160008	1 316	1 316	0
EU3	160888		0	1 328

There is no problem here with the matching key, BIN. All BIN numbers are correct and the related legal units have been matched without any errors – no false matches and no false non-matches. However, the legal units are not suitable as statistical units. We have the same population of legal units generating the data, but many of these legal units are related to other legal units through ownership patterns.

Behind each legal unit in Chart 4.8a there is a 'family' of related legal units. The person at the legal unit 160006 received a questionnaire from Statistics Sweden and responded for the whole 'family'. But the tax reporting within the 'family' was handled by another legal unit (160666), which is clear from Chart 4.8b.

Three different populations consisting of different kinds of units are behind the legal units shown in Chart 4.8:

– Those that received the questionnaire – *Statistics Sweden's reporting units*

– Those that report yearly income tax returns – *Income tax return units*

– Those that report VAT – *VAT return units*

When we match these sources we obtain correct matches of legal units, but statistically erroneous matches. This is one of the most difficult issues in economic statistics: How to achieve consistent estimates from different surveys?

Legal units belonging to big business are almost always included in this kind of enterprise 'families', or to use the correct term, they almost always belong to *composite enterprise units* together with other legal units. However, many small legal units can also be related to other small legal units in similar ways.

Statistical units have changed

In this example, the Annual Pay Register is compared with the Quarterly Pay Register. These registers are used by the yearly and quarterly National Accounts for estimating wages and salaries by institutional sector and economic activity. The Quarterly Pay Register is based on monthly reports from all employers delivered about two weeks after the end of each month. The Annual Pay Register is based on reports from all employers delivered about one month after the end of the year.

If we match the monthly sources and the yearly source with the matching key BIN, we find patterns that are illustrated in Chart 4.9. Again, the problem is not the record linkage as the legal units have been combined correctly. The problem is that the units have changed – there has been one takeover, the legal unit BIN1 has taken over the legal unit BIN2.

Chart 4.9 Comparing gross yearly pay in quarterly and annual registers

BIN	ISIC	Quarterly	Annual	
BIN1	29	25	110	The legal units BIN1 and BIN2 have merged into BIN1
BIN2	25	84	0	

The legal units that merge have in many cases different ISIC codes. This fact will generate inconsistencies between the quarterly National Accounts that use the Quarterly Pay Register and the yearly National Accounts that use the Annual Pay Register.

4.7   The statistical system and record linkage

What happens when a country with a traditional census-based statistical system starts to use administrative registers? Identities are not used in a traditional statistical system. The frames used for sample surveys and censuses are area-based, and combining microdata from different surveys is not necessary. All variables required for a survey are collected during the interview, and the interviewers do not ask for national identity numbers. This means that record linkage is not practised.

When the transition into a register-based system starts, one of the first projects will be the creation of a statistical population register. Evaluating the quality of the statistical population register will require comparing the register with the most recent census. The register contains identity numbers, but the census usually has no identity numbers. Therefore, record linkage must be conducted using names, birth dates, etc. as matching keys. This means that bad quality identities must be compensated with complicated linkage methods – probabilistic methods must be used in many cases.

If there will be a new census in the future, the interviewers should ask for the identity numbers of the interviewed persons. The interviewers should also ask for identity numbers in all area samples of households. This is necessary for quality assurance and opens new possibilities for quality improvements, for example, with calibration estimators.

Since area sampling is a very costly survey method, the new statistical population register will be used as sampling frame in the future. Then requesting identity numbers will not be necessary – that information is already in the sampling frame. Telephone interviews with persons selected from a population register will now become a data collection method that the NSO can use.

The conclusion is that when official statistics becomes register-based, survey methods will change. Record linkage is one important survey method that will change completely. Low-quality identity variables used for difficult probabilistic record linkage will be replaced by national identity numbers and efficient high-quality deterministic record linkage.

Registers for research – short-term and long-term strategies

We had contacts with a research institute in Latin America that uses several registers with data for many millions of persons. Record linkage with social security numbers resulted in research registers with combined data from a number of social statistics registers. The institute also wanted to combine this information with health register data from hospitals. However, the hospitals have their own system of identity numbers and do not use social security numbers.

Short-term strategy: Use advanced probabilistic record linkage to combine the social statistics data with the health data. This is the strategy that the research institute had chosen.

Long-term strategy: Improve the statistical system so that the hospitals also use the social security numbers. Develop the system for social security numbers into a system with national identity numbers for the entire population. Use deterministic record linkage with identity numbers to combine the social statistics data with the health data. This is the strategy that the institute should choose.

CHAPTER 5

Building the System – Quality Assessment

Quality assessment of administrative registers is undertaken when an National Statistical Office decides on how to treat new potential sources. Moreover, the statistical registers created with these sources must be evaluated before the new register-based statistics are published. Established principles for quality assessment in a census-based statistical system with sample surveys may be unsuitable when an NSO starts to work with register-based statistics. Some basic concepts and principles are presented here that are relevant for the work with registers.

This chapter discusses the inventory and evaluation of administrative sources and introduces a number of quality indicators. Related topics, such as editing of each source and comparisons with other sources, are discussed in Chapter 6.

5.1 Four quality concepts

The production process for a register survey, where microdata from administrative and statistical registers are used to create a new statistical register, has been described earlier. Chart 5.1 introduces four different quality concepts that are relevant for different parts of the production process.

Chart 5.1 A register survey: From administrative registers to statistical estimates

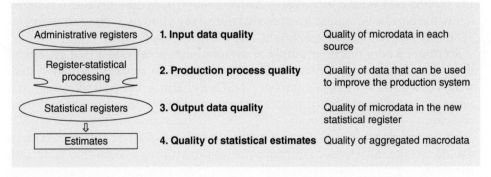

Administrative registers	1. Input data quality	Quality of microdata in each source
Register-statistical processing	2. Production process quality	Quality of data that can be used to improve the production system
Statistical registers	3. Output data quality	Quality of microdata in the new statistical register
Estimates	4. Quality of statistical estimates	Quality of aggregated macrodata

Register-based Statistics: Registers and the National Statistical System, Third Edition. Anders Wallgren and Britt Wallgren.
© 2022 John Wiley & Sons Ltd. Published 2022 by John Wiley & Sons Ltd.

When a new statistical register is created, input data can consist of one or several administrative registers and/or already existing statistical registers. The first step in the work with creating a new statistical register consists of an inventory of potential sources. Thereafter the work with quality assessment starts.

Two serious and deeply rooted misunderstandings will lead to the administrative registers not being evaluated and used as they should.

The "one survey at a time" paradigm

In the traditional statistical system, treating each sample survey individually is sufficient. The microdata collected for the survey are meant to be used only for one purpose. The survey will not be combined with other sample surveys.

If the same paradigm is used when an NSO starts work with administrative sources, then many opportunities and benefits from working with registers will be missed. One and the same administrative source can be used in many ways, which must be understood when the source is evaluated. Quality, or *fitness of use*, depends on how we use the source. Even if a source is bad for one specific use, it could be good for another use.

Example: An authority is responsible for the support of poor people and maintains a register of all support recipients. The register covers about 80% of the population. However, the *input data quality* for population statistics is bad due to selective undercoverage. But the *production process quality* is high – the register has up-to-date information on residential addresses that can be used to improve the Population Register, which is an important part of the statistics production system.

"We have no control over the collection of administrative data"

As Laitila, Wallgren and Wallgren (2012) state: *In traditional sample surveys the NSO is in control of the data collection and can evaluate the quality of the statistics produced …. The situation is different using register surveys since the NSO has no control over the data collection and registration phase.*

We have also been guilty of this misunderstanding. But as noted in Chapter 2, the Nordic NSOs have been clearly active in improving administrative systems. Thus, the Nordic administrative registers now have good quality information on *Where do people live, and with whom, and where do people work?*

Section 2.3 contains an example of an administrative system that generated dirty data for criminal statistics. The solution was to replace the bad system with a new good system. A number of NSOs in Latin America understand that parts of the national administrative systems must be improved and have started reforms.

The conclusion is that once a source has been analysed, the NSO can decide to take initiatives to improve the administrative system that generates the data.

5.2 Making an inventory of potential sources

An inventory should be made of which sources could be used to create new statistical registers. There may be hundreds of authorities in a country that have created administrative registers and new systems are being developed all the time. Thus, the inventory should cover a large and complicated area. Many new partners will be found and cooperation with some of them must be established.

This inventory should be done by subject matter specialists at the NSO. Working groups can be organised, and each group should have a defined area. For example, one group can search for administrative data that could be used for environmental statistics; another group could search for sources for population statistics,[1] etc. Chart 5.2 contains a number of registers that could be among the first to be considered. The inventory should be carried out step by step, where different working groups search for sources for a specific register. A statistical population register and a business register are registers that should be considered early.

Chart 5.2   Statistical registers that are suitable for early development

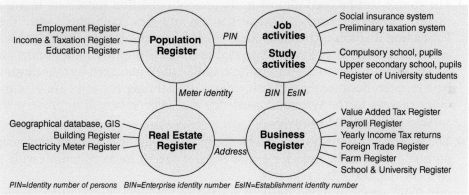

PIN=Identity number of persons   BIN=Enterprise identity number   EsIN=Establishment identity number

During inventory and source evaluations, a number of working groups from the NSO will acquire *metadata* regarding several administrative registers. This metadata should be stored in a metadata system for input data, as described in Section 2.4.2.

5.3 How can a source be used?

All available sources with a connection to the problem area should be analysed when a new register is created. Every new source could potentially be used to improve coverage in the new register regarding objects and variables. A new

---

[1] Dias et al. (2016) describe an inventory of sources for a register-based census in Portugal.

source can also mean that inconsistencies are detected which could contribute to improved quality. The work with quality assessment of an administrative register can result in one of the following decisions:

1. Do not use the source at present. The administrative system must be improved before it can be used in the statistical production system. Do not spend time on dirty data from authorities that are difficult to work with.

2. The source can be used independently. If the administrative object set and administrative variables are of good statistical quality, the source can be used independently for producing statistics, even if there are no good matching keys.

   This usage of administrative sources will be the first kind of register survey design used in new register countries that lack a statistical register system with base registers. The statistical methods used will involve micro-editing of the source to find and correct unreasonable values; and macro-editing to compare estimates based on the source with estimates from other surveys. New derived variables can also be created using the administrative variables.

3. The source can be used almost in its present condition for a new register survey. However, nearly all new sources are first combined with the Population Register or Business Register. The benefits derived from combining a source with the relevant base register include improved knowledge on coverage and access to the important classifications managed by the relevant base register. Because these classifications or spanning variables are used by other surveys in the system, consistency and coherence will improve – all surveys use the same variables.

4. The source has some quality problems, but after suitable register-statistical processing, it can be used for a new register survey in combination with a base register and other registers.

   In many cases, different parts of the desired population use different administrative systems. For example, small enterprises report in one way using one set of tax forms, while larger enterprises report more often and with more variables. Many administrative registers must be combined to obtain the necessary coverage in these cases.

   In addition, different variables may be reported in different administrative systems. For example, wages are reported in one system, pensions in another, economic support to poor families in a third, etc. Then combining many administrative registers into one statistical register becomes necessary to gain a complete picture of the incomes of persons and households.

   In many cases, the administrative system is used for one specific category and must be combined with the base register in question to enable comparisons. For example, the University Register must be combined with the Population Register to enable comparisons between young people who study and those who do not study. In these cases, it will be important to import variables from other registers in the system, in addition to variables from the base register.

5. The source is already used for a register survey, but if it is combined with other sources a new more advanced register survey can be designed.

6. The source can be used to improve the production system and the quality of already existing sample surveys and register surveys.

A new statistical register will always improve the system. As all registers and all sample surveys can be combined with the new register, the variable content will be improved and new combinations will be possible. When working with the new source, comparisons are made with a base register and other registers and sample surveys in the system. New kinds of errors can be found and corrected.

Sample surveys and censuses are carried out with *one* particular use in mind and quality issues generally focus on the estimates. In the case of a statistical register, *many* different uses are possible – it may serve not only current surveys but also future ones.

Similar to other surveys, the quality of a *register survey* also relates to *one* specific use of the register. It focuses on the quality of the estimates, particularly their relevance and accuracy in relation to the survey's purpose. Here, describing quality is a question of indicating whether the quality of the survey is *good or bad*.

However, the quality of an *administrative or statistical register* is not related to only one particular use. Indicating the register's *characteristics* is important when describing quality in this respect, thereby implying the uses to which the register may be put. The quality of the register will affect the quality of the surveys based upon it, and is determined by three factors:

– The administrative systems that generate the input data.

– The possibilities offered by the system of statistical registers with regard to improving coverage, content of variables and consistency.

– The processing methods used to produce the register.

In Chart 5.3, three administrative sources are used to create seven statistical registers. The output quality of the estimates produced with these seven registers depends on the input data quality of the three sources; the possibilities offered by the register system; and the processing methods used when creating these seven statistical registers. Chart 5.3 illustrates these aspects.

Chart 5.3 Input and output data and the production process

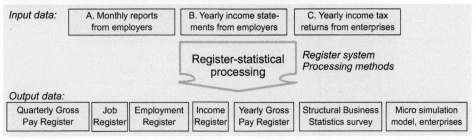

5.4   Quality assessment in a register-based production system

Different kinds of survey errors are utilised as planning criteria in the work with survey design. This is well known and discussed for the design of sample surveys.

How should the corresponding planning process for register surveys be structured? In Laitila, Wallgren and Wallgren (2012), we describe the systems approach to survey design as consisting of four steps A–D as illustrated in Chart 5.4. These four steps have four different groups of quality indicators. Quality assessment is done not only to find signs of bad quality – it also aims to find opportunities and understand how a source should be treated. Each administrative source is analysed as follows:

A. Metadata regarding the source are analysed. The relevance of input data quality is determined as described in Section 5.4.1.
B. Microdata from the source are analysed. Different aspects of the accuracy of the input data quality are determined as described in Section 5.4.2.
C. The source is compared with its base register. Some aspects of accuracy of the source and the base register are determined. A decision is made regarding if the source can be used to improve the base register. Input data quality and production process quality are analysed. This is described in Section 5.4.3.
D. The source is compared with all surveys in the system containing *similar variables*. Aspects of the accuracy of the source and the surveys used for comparisons are determined. Other decisions concern whether the source can be combined with other sources for a new survey and whether the source can be used to improve other surveys. Input data quality and production process quality are analysed. This kind of editing is described in Section 5.4.4.

Each administrative register should be analysed to see if it is usable for the survey in question. Daas et al. (2011), Daas, Ossen and Tennekes (2012) and Laitila, Wallgren and Wallgren (2012) describe quality indicators that can be used for quality assessment of administrative registers. When new sources are analysed to see if they could be used for a new statistical register, this decision can be based on the indicators in these reports.

Chart 5.4   The work with quality assessment of an administrative source

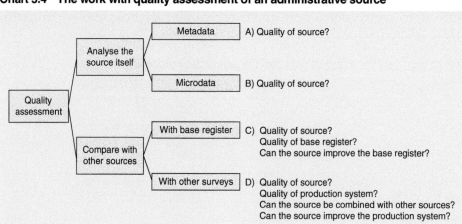

The indicators presented below are an extended version of the indicators in Laitila, Wallgren and Wallgren (2012).

As a rule, the understanding of how an administrative source should best be used by a statistical office requires time to develop. New competence and new methods must be developed. Since there may be many potential ways of using the source, evaluating the source more than once may be necessary. Each administrative register or source that a statistical office considers using for statistical purposes should first be analysed to see if it is usable and how it could be used.

The quality indicators in Charts 5.5–5.8 are used for the quality assessment of an administrative source. During the work with these indicators, the statistical usability of the source is analysed; many usages are considered; and many combinations with other sources or surveys are evaluated. Section 5.6 presents a fifth group of indicators that describe the quality of the base registers in the register system.

5.4.1 Analysing metadata

The first set of indicators in Chart 5.5 is based on the analysis of metadata information from the administrative authority responsible for the source. Tax forms, supporting brochures, handbooks, etc. should be studied. This is the first step in the work of analysing an administrative register or source. Interviewing persons at the administrative authority responsible for the source is also recommended.

Chart 5.5   Indicators A1–A9 of input data quality – Relevance

Indicator	Quality factor	Description
A1	Relevance of population	Definition of the administrative object set. Which administrative rules determine which objects are included? Is this set suitable as a statistical population?
A2	Relevance of units	Definition of the administrative units. Are these units suitable as statistical units?
A3	Relevant matching keys, identity and reference variables	Are there suitable identity and reference variables in the source for micro integration within the NSO? Are there identity numbers, names, birthdates, and places of birth and addresses that can be used as matching keys?
A4	Relevance of variables	Definitions of the administrative variables. Are these suitable as statistical variables?
A5	Relevance of reference time	Are reference times suitable for statistical usage? What rules are used for accruing accounting data between months and years?
A6	Study domains	Can the units be allocated between relevant study domains? Are there variables describing domains in the source, or can the units be linked with domain variables in the Base Register?
A7	Comprehensiveness	Does the source contain a small/large part of an intended population? Does the source contain few/many statistically interesting variables? Can a small/large number of existing surveys benefit from the administrative source?
A8	Updates, delivery time and punctuality	How often and at what points in time is the administrative register updated? Time for delivery of the source from register holder to the NSO. Difference in time between delivery and agreed delivery time.
A9	Comparability over time	Extent of changes in the content of the administrative register over time.

5.4.2   Analysis and data editing of the source

The indicators in Chart 5.6 are based on analysis and editing of the source microdata. The usual statistical description and exploratory data analysis should be performed. After that, an analysis resembling usual editing should be performed. The quality of matching keys and time references are very important, as pointed out by Cain, Figueroa and Herrera (2019). Matching keys and time references are very important variables for register surveys. Traditional system surveys do not use matching keys, and time references are determined in the questionnaire.

The aim of this analysis is to diagnose the source and should not be confused with the editing performed during routine production of statistics. A clear understanding of the administrative variables is necessary to create good editing rules. Comparing preliminary and final data is also important when taxpayers deliver corrected and delayed tax reports.

Chart 5.6   Indicators B1–B7 of input data quality – Accuracy

Indicator	Quality factor	Description
B1	Quality of identifying variables: Primary keys Is it possible to link identical units?	Fraction of units with usable identifying variables. The quality of all variables to be used as matching keys is described. a) Fraction of units with usable *identity numbers*. The identity numbers should have correct format and reasonable values. b) Fraction of units where *names* have correct spelling and format. c) Fraction of units with usable *birth dates* and *birthplaces*. d) Fraction of units with usable *addresses*.
B2	Quality of reference variables	Is it possible to link with related units? Fraction of units with usable reference variables (foreign keys).
B3	Duplicates	Fraction of duplicates with the same or almost the same identifying variable values.
B4	Time references	Fraction of observations where time references are missing.
B5	Missing values	Fraction of missing values for the variables of statistical interest.
B6	Wrong values	Fraction of wrong or unreasonable values for the statistically interesting variables.
B7	Quality of preliminary data	Fraction of observations corrected by the taxpayers. Estimates based on preliminary data are compared with estimates based on final data.

5.4.3   Comparing a source with the base register

The integration with the base register will result in three object sets: units only in the source, units only in the base register, and units in both. What does the non-match indicate? Are there quality problems in the base register or in the source? The statistical units that constitute the non-match between the source and the base register should be analysed carefully. Investigating whether the source can be used to improve the quality of the base register is also important.

Chart 5.7   Indicators C1–C5 of accuracy when comparing with the base register

Indicator	Quality factor	Description
C1	Undercoverage in base register	Measures production process quality. Fraction of units: Are there units that have been active during the reference period but are missing or coded as inactive in the base register?
C2	Overcoverage in base register	Measures production process quality. Fraction of units: Are units coded as active in the base register and belong to a category that is covered by the source, but they have no reported activity in the source?
C3	Undercoverage in the source	Measures input data quality. Fraction of units: Are there units that have been active during the reference period according to the base register but are missing in the source?
C4	Overcoverage in the source	Measures input data quality. Fraction of units: There are units in the source that belong to a category, or seem to belong to a category, that is not statistically relevant.
C5	Can the source improve base register?	Measures production process quality. A more thorough analysis is required here depending on the character of the source. The quality improvements should be measured.

To distinguish between C2 or C3 can be difficult, the symptoms are similar. Subject matter competence is required to make the correct interpretation. According to our experience the problems are often in the base register.

5.4.4   Comparing a source with surveys with similar variables

The source should be integrated with relevant register surveys and sample surveys with variables that are similar to those in the source. What characteristics does the source have? How can it be used? There may be many ways of using a source. How should the source be treated to make it usable? Should it be combined with other sources? The following quality issues are important here:

1. When the source is compared with other surveys, do we find errors in the source? This is indicator D1.
2. When the source is compared with other surveys, do we find errors in some of the other surveys? This is indicator D2.
3. Can the source be used to improve other surveys, or the production system? Can the source by used to reduce errors? This is indicator D3.
4. Can the source be combined with other surveys so that the combination can be a new survey? This is indicator D4.

Chart 5.8   Indicators D1–D4 of accuracy when comparing with related surveys

Indicator	Quality factor	Description	
D1	Is the source good or bad?	a) Compare populations b) Compare units c) Compare variables	Measures input data quality
D2	Are the related surveys good or bad?	a) Compare populations b) Compare units c) Compare variables	Measures production process quality of the source and output data quality of related surveys
D3	Can the source improve other surveys?	a) Will populations be better? b) Will units be better? c) Will variables be better?	Measures production process quality
D4	Can the source be combined with other sources?	a) Will population be better? b) Will units be better? c) Will variables be better?	Measures input data quality

Analysing these issues is an important and difficult task that requires very competent staff with a good overview and familiarity of the production system. Subject matter competence and familiarity with statistical analysis are important.

5.5   Output data quality and quality of estimates

Assume that an NSO has created the first version of a statistical population register. Can that register be used to produce regional population statistics? All relevant administrative registers were used when the population register was created. This means that quality assessment of the new register should be done using comparisons with other sources. The new register can be evaluated through three different kinds of surveys:

- A test census in one or some selected regions.
- A sample from the new register. If the register includes telephone numbers, data can be collected via telephone interviews.
- An area sample with interviewers who conduct face-to-face interviews.

5.5.1   Analysing quality with a test census

In Ecuador, a test with a traditional census including identity numbers was carried out on the Galapagos Islands during 2015. This census was compared with available administrative registers for Galapagos. Requesting the identity numbers of the persons in the census gave no problems and the identity numbers had good quality. Hence, deterministic record linkage could be done with the population register.[2]

---

[2] Based on information from our colleagues at Unidad de Metodología de Registros Administrativos, Dirección de Registros Administrativos, Instituto National de Estadística y Censos (INEC), Ecuador.

Chart 5.9 Coverage errors in the population register for Galapagos 2015

Test census 2015 for Galapagos		Population Register for Galapagos 2015	
Population size, N(census)	25 100	Population size, N(register)	30 373
Deterministic matching with register Matching key: Identity number	17 830	Deterministic matching with census Matching key: Identity number	17 830
Probabilistic matching with register Matching key: Names, 85% similarity	67	Probabilistic matching with census Matching key: Names, 85% similarity	67
Undercoverage in the register: persons in the census, but not in the register	7 203	Overcoverage in the register: persons in the register, but not in the census	12 476

The results in Chart 5.9 show that the population register for Galapagos is not suitable for statistical purposes – the coverage errors are serious. People leave or move to Galapagos without reporting to the office responsible for the register. Migration in Ecuador is not recorded with sufficient quality with the present system for national registration. Thus, regional population statistics become very difficult. That is why the country is planning to develop the *Register of Residence*, as noted in Section 2.7.1.

5.5.2 Analysing quality with samples from the new register
The statistical population register is the most important part of the national statistical system and the basis for all social statistics. Therefore, the quality of this register must be analysed carefully. The coverage of the population register determines the coverage of all registers with data on persons and all sample surveys of persons and households. The *coverage* should be evaluated preferably with samples based on area frames. The quality of the *variables* in the population register can be evaluated with both register-based and area samples.

The quality of the variables in the new population register can be analysed by taking a sample from the register. The sample will give information valid for the whole country, unlike a test census that gives information for only one or some selected regions. There are variables in the register that can be used to contact persons. The sample survey will be easy to conduct if these contact variables include telephone numbers. The quality of these central variables can be analysed by comparisons with the register information for each person in the sample based on identity numbers from the register and answers to the following central questions:

– Where do you live and with whom? In which municipality do you live?

– Do you work? Where do you work? In which municipality?

These questions are part of a normal Labour Force Survey (LFS) interview. The LFS-survey can be used in this way to monitor the quality of the new population register as well as a new Employment Register.

An example with data from population registers

We illustrate some quality concepts with microdata from the Swedish Population Register. We use linked microdata from Statistics Sweden's Population Register for 2005 and 2015 regarding persons who belonged to the population both these years. This means that the only demographic events we study are migration within Sweden. We believe that this is the main quality issue for new register countries.

The register for 2005 is the register that is evaluated and will be regarded as a badly updated population register for 2015. This register will be compared with the 'true' values based on data from 2015. In a real application, we do not know the true values. We will use this set of data as test data for comparing different estimation methods in Section 8.3.

Output data quality of microdata regarding the variable *Home municipality*: 79.3% of the population have the correct home municipality.

Quality of estimates can be measured with absolute errors. The register could be used to estimate the population size for each of the 290 municipalities. The quality of these estimates can be described by absolute errors. The table below describes the quality of estimated population size for two municipalities.

One rural municipality: Population estimate: 2 700		Absolute error:
Overcoverage due to people have moved out:	619 persons	$100 \cdot |2700\text{-}2386| / 2386 =$
Correct municipality:	2 081 persons	
Undercoverage due to people have moved in:	305 persons	$= 6.6\,\%$

One urban municipality: Population estimate: 670 474		
Overcoverage due to people have moved out:	145 521 persons	$100 \cdot |670474\text{-}718008| / 718008 =$
Correct municipality:	524 953 persons	
Undercoverage due to people have moved in:	193 055 persons	$= 13.2\,\%$

The common approach is to use register data to calibrate the sampling weights in sample surveys with nonresponse errors, with the aim to reduce these errors. Here we work in the opposite direction: we use sample survey data to calibrate weights in a register with coverage errors with the aim to reduce the coverage errors. The table below illustrates the possibility of reducing such errors with the calibration technique. We return to this example in Section 8.3 and compare estimators.

290 population estimates	Mean absolute error, %	3rd quartile	Maximum absolute error %
Population register, no calibration	4.4	6.4	16.8
Population register, calibrated weights	0.6	0.6	6.0

5.5.3   Analysing quality with area samples

A statistical register with coverage errors due to undercoverage can be combined with a sample survey based on an area frame. Since samples based on area frames theoretically have no coverage errors, the estimates can be unbiased.

Combining a register with an area-based sample becomes necessary with coverage errors and problems with residential addresses that have not been updated. Producing estimates by municipality is desirable with the register. The area-based sample is used to *estimate coverage errors* on the national and regional levels as well as for categories of municipalities and persons or households. The area sample cannot be used to produce estimates for municipalities, but it can be used to produce estimates for the smaller number of categories. *Adjustment for coverage errors* is also possible with calibration conditions based on register variables.

Auxiliary variables

Assume that we want to produce statistics for municipalities with the Population Register created by the NSO. The municipalities have been classified in different categories based on the degree of urbanisation and perhaps the relative size of an indigenous population. We believe these variables are correlated with the coverage errors for municipalities. In addition, classifying households or persons in categories with different propensities of moving to another address may be possible with such variables as age, level of education and economic activity. These categories of municipalities and/or persons are the auxiliary variables that can be used to reduce coverage errors in the register.

For an area sample, interviewers collect the same variables as in the register. We will find persons in the sample who are not in the register or have a different address, municipality or district in the register. We will thereby find undercoverage and overcoverage on different regional levels.

Estimation of coverage errors

A number of important quality measures can be calculated based on comparisons between the area sample and the register: percentages of persons with wrong residential address as well as under- and overcoverage rates on the regional and national levels. The long-term work of improvements of the national registration system should be monitored with these measures.

Chart 5.10 illustrates different kinds of coverage errors. On the national level, the area sample gives an estimated undercoverage of 72 000 persons in the register. The area sample estimate is that 80 000 persons are actually living in district A, but they are registered as living in B. These 80 000 persons are the estimated undercoverage of the subpopulation living in district A and at the same time an estimated overcoverage of the subpopulation living in district B.

Persons who have left the country is another category of overcoverage that cannot be detected with the methods discussed here, but the interviewers can ask if someone in the household has emigrated.

Chart 5.10   Coverage errors and area sample estimates of the population by district

		District according to register data			Not in		Under-
	District	A	B	C	register	All	coverage
District	A	2 870 000	80 000	110 000	19 000	3 079 000	209 000
according to	B	16 000	3 110 000	40 000	27 000	3 193 000	83 000
area sample	C	2 000	10 000	1 530 000	26 000	1 568 000	38 000
	All	2 888 000	3 200 000	1 680 000	72 000	7 840 000	330 000
	Overcoverage	18 000	90 000	150 000			

5.5.4   Measuring quality of basic register variables with the LFS

The national statistical system should be designed so that the following questions can be answered with estimates from statistical registers: *Where do people live, and with whom? Where do people work?* This is discussed in Sections 2.6–2.8. The quality of these important variables should be measured every year.

If the Labour Force Survey includes identity numbers of persons, the information in the LFS can be compared with corresponding information in the statistical registers: Do people live where we believe? Are the households as we believe? Do people work where we believe? The quality of the Population and Employment Registers are thereby verified.

5.6   A coordinated system of registers

The *System principle* in Chart 1.13 notes that all statistical registers should be included in a coordinated register system. The *Consistency principle* is also noted, stating that consistency regarding populations and variables is necessary for the coherence of estimates from different statistical registers. If the register system follows these principles, *the system is coordinated*, which means that registers can be combined without disturbing mismatch. Here, we concentrate on the importance of coordinated populations and how lack of coordination can be measured.

5.6.1   Are the base registers a coordinated system?

The register system in Chart 5.11 consists of four parts defined by the base registers. The registers in each part contain statistical *units of the same kind*. The consistencies within each part are measured with the quality indicators C1–C5 in Chart 5.7. The four base registers contain statistical *units of different kinds* and consistencies between them should be monitored by other quality indicators:

a. Can all persons in the Population Register be georeferenced? How many persons have a valid meter identity (a proxy for dwelling identity) that is included in the Electricity Meter Register?
b. Are there persons in the Activity Register with job activities and study activities that are not included in the Population Register? This will generate inconsistencies between social statistics and economic statistics. The frame population for a register-based Labour force Survey, LFS, will not include the whole labour market.
c. Are there enterprises and establishments in the Activity Register that are not included in the Business Register? This will generate inconsistencies between labour market statistics and business statistics. Hours worked in a register-based LFS will not be coherent with production in the Structural Business Statistics survey, SBS.
d. Can all establishments in the Business Register be georeferenced? How many establishments have an address that can be linked with the Real Estate Register?

The quality indicators above are relevant when we want to link different kinds of statistical units in different base registers: Persons – Dwellings (a), Activities – Persons (b), Activities – Establishments (c), and Establishments – Addresses (d).

This kind of advanced linking is necessary for the registers marked with numbers in Chart 5.11. Many derived variables are created by *aggregation* as described in Section 4.4.

– The Employment Register (1) is created mainly by aggregating variables from the Social insurance system (4).

– The Income & Taxation Register (2) is created by aggregating variables from the Taxation system (5).

– The Education Register (3) with level and kind of education of persons is created by aggregating variables from the registers with the activities of pupils and students (6).

– The Payroll Register (7) with wage sums is created by aggregating variables from the Taxation system (5).

– The School & University Register (8) with 'production' of education is created by aggregating variables in the registers with the activities of pupils and students (6).

Chart 5.11   Creating registers by linking different kinds of statistical units

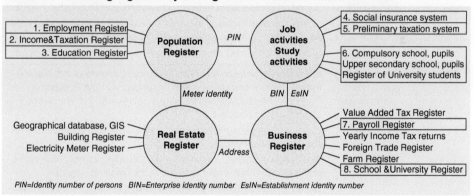

PIN=Identity number of persons   BIN=Enterprise identity number   EsIN=Establishment identity number

5.6.2   Quality indicators at the system level

The four parts of the register system in Chart 5.11 are based on different kinds of administrative systems regarding different kinds of units. That is why consistency between populations from different parts needs special attention. For example, if we link activity data from the registers numbered 6 in the chart, a number of inconsistencies will give rise to quality problems. These problems will result in many cases with *non-match* due to differences in *coverage*, when we combine registers with record linkage.

– Some pupils and students in 6 are not in the Population Register.

– Some school and university units in 6 are not in the register numbered 8. Also, inconsistencies can exist between the School & University Register (8) and the Business Register.

– Similar problems will arise when we link job activity data in 4 and 5 with the Population Register. Some persons with jobs as employed are not in the Population register. And some establishments in 4 and 5 may not be in the Business Register.

In contrast with these problems, the registers numbered 1, 2 and 3 are perfectly consistent with the Population Register (See Chart 1.14). But the populations in the group with the Business Register are all inconsistent because the administrative systems differ. And finally, the agricultural populations in the Business Register and the Farm Register can be inconsistent as pointed out in Wallgren and Wallgren (2010).

Chart 5.12 presents important quality indicators regarding the register system intended to be the basis of the national statistical system. Inconsistencies between base registers and the three basic census issues (*Where do people live, and with whom? Where do people work?*) are measured with these indicators. Our impression is that these quality problems are often forgotten.

Chart 5.12   Indicators E1–E4 of the quality of base registers in the system

Indicator	Quality factor	Description
E1	Consistency between populations in the base registers	a) Fraction of persons who are included in the Activity Register but are missing in the Population Register.
		Based on comparisons between the Activity and Population Registers.
		b) Fraction of enterprises and establishments that are included in the Activity Register but are missing in the Business Register.
		Based on comparisons between the Activity and Business Registers.
E2	Quality of geo-referencing of persons	a) Fraction of persons for which measured geocode of the dwelling (or the electricity meter) agrees with the geocode in the Population Register. The geocode can be Region, Municipality or Home address, etc. *'Where do people live?'*
		Based on comparisons between an area sample or a register-based sample and the Population Register.
		b) Fraction of persons who have a valid dwelling identity (electricity meter identity) that is also included in the Dwelling Register.
		Based on comparisons between the Dwelling Register (electricity meter register) and the Population Register.
E3	Quality of dwelling households	Fraction of households where the measured household in the LFS agrees with the household in the Population Register *'With whom do people live?'*
		Based on comparisons between an area sample or a register-based sample and the Population Register.
E4	Quality of geo-referencing of jobs	a) Fraction of employed where the establishment in the LFS agrees with the Employment Register. *'Where do people work?'*
		Based on comparisons between the Labour Force Survey and the Employment Register.
		b) Fraction of establishments in the Business Register that have a valid address that can be linked with the Real Estate Register.
		Based on comparisons between the Real Estate and Business Registers.

5.7   Using the quality indicators

In Laitila, Wallgren and Wallgren (2012), the indicators in groups A–D are used to analyse the yearly *income statements* from all employers regarding gross pay, social security contributions and preliminary tax during 2009. Individual data regarding all employees are delivered during January after the year in question. About 60% of total taxes in Sweden are collected with this system. The Tax Board receives most of this data online or as data files submitted by employers.

A. Metadata – information from the Administrative Authority
The relevance of income statements for statistical purposes should be assessed by subject-matter specialists. As this is work with economic statistics, experts from the unit working with the National Accounts should be consulted. The tax form with explanations and the brochure on income statements (about 40 pages) available for all employers is the main source of information from the Tax Board that should be analysed. The results of this analysis regarding the income statements are summarised in Chart 5.13.

Chart 5.13   Information from the administrative authority – relevance

Indicator	Quality factor	Description
A1	Relevance of population	The source contains information on jobs as employed, employed persons, enterprises that are employers, and establishments where employed persons work. All are relevant as statistical populations.
A2	Relevance of units	The source contains four kinds of relevant units (jobs, employees, enterprises that are employers, establishments with employees).
A3	Relevant matching keys	Three important keys are combined: Identity number of the employer, Personal Identity Number of the employee, and Establishment number.
A4	Relevance of variables	Gross salary on the tax form plus benefits correspond to the definitions used by the National Accounts.
A5	Relevance of reference time	The Income Statements give information on wages and salaries paid to the employees during the calendar year. This definition is in accordance with the requirements of the National Accounts.
A6	Study domains	All kinds of study domains are possible using classification variables from the Population Register or Business Register.
A7	Comprehensiveness	The source covers all employees and all employers. The source is comprehensive.
A8	Updates, delivery, punctuality	The source is yearly. Income statements are delivered to the Tax Board during January, but corrections are made during the whole year. Preliminary statistics can be produced before summer and final estimates during the autumn.
A9	Comparability over time	Comparability over time is good.

The relevance of this source is very high. This data source is necessary for the Employment Register that is a part of the register-based census. The income statements are also the best source for statistics on gross wages and are used by the National Accounts. Since three identities are combined in the income statements, this source is a very important part of Statistics Sweden's production system that enables linkage of records from many different sources.

B. Analysis and Data Editing of the Source

The data in the income statements are usually generated by the employers' computer systems or by an internet-based application. This means that editing is done at the same time as the data are generated. Identities must be correct, as tax payments of persons and enterprises are administrated with income statements.

However, two variables in the income statements are not used by the Tax Board but are used only by Statistics Sweden. These are *employment time* and the *work site number* on the income statement. If these variables are analysed, statistically important quality issues are found. These findings are described in Chart 5.14.

Chart 5.14   Information from analysis and data editing of the source – accuracy

Indicator	Quality factor	Description
B1	Quality of identifying variables (Primary keys)	190 701 or 6.4% of all income statements from enterprises with more than one establishment have missing establishment/ establishment identities.
		After a register maintenance survey of about 4 400 of these enterprises, the establishment identity on 188 962 income statements was corrected.
B2	Quality of reference variables	Link to the Population Register – PIN usable: Of employed persons, 5 028 405 or 99.94% had a usable PIN; 3 107 did not have a usable PIN.
		Link to Business Register – BIN usable: All
B3	Duplicates	Duplicates are not a problem.
B4	Time references	Quality of the variable employment time has low quality, see B6 below.
B5	Missing values	Employment period defined as the month from, and month up to. 0.06% values are missing.
B6	Wrong values	Employment time defined as the month from, and month up to. Many employers answer from January up to December even if the actual work was done during a shorter period. Aggregate wage can be small, but employment period can be 'long'. This indicates measurement errors.
B7	Quality of preliminary data	Income statements are corrected by employers and this causes delay. Preliminary and final estimates were compared. The decision was made that early estimates based on data available during September should be used instead of the final data in December.

Accuracy is generally good, but the input data quality of the establishment identity numbers is not sufficiently good. However, after a *register maintenance survey* where questionnaires were sent to more than 4 000 enterprises, the quality of this variable is now sufficiently good.

Note: Register maintenance surveys are sometimes an important method to improve quality. For a specific category of units, we suspect that quality is bad. Then we can send questionnaires or conduct telephone interviews for this category. If there are units with missing values for important variables, questionnaires can be sent or telephone interviews can be conducted also for this category.

C. Integrate the Source with the Base Register

The Income Statement Register is an important source for the Activity Register, which is one of the four base registers. Income statements can be linked with two other base registers – the Population Register and the Business Register. Income statements that cannot be linked with these base registers indicate undercoverage in these registers. The quality indicators C1–C5 in Chart 5.15 are based on comparisons between the Income Statement Register and the base registers.

Chart 5.15   Information from integrating the source with base registers – accuracy

Indicator	Quality factor	Description
C1	Undercoverage in base register	In all, 57 905 foreigners who work and pay tax in Sweden were found in the Income Statement Register but were not found in the Population Register. The fraction of undercoverage among the population of all employed persons in the Employment Register is 1.4%.
		There were 315 380 enterprises in the Business Register that were classified as active employers during one calendar year. According to the income statements, there were 31 393 more enterprises active as employers during the year in question.
C2	Overcoverage in base register	Of the 315 380 enterprises that were classified as active employers, 11 301 enterprises, or 4%, were not active according to income statements.
C3	Undercoverage in the source	Black work is a problem.
C4	Overcoverage in the source	No problem
C5	Can the source improve the base register?	Comparisons with the Income Statement Register show that the coverage of both the Population and Business Registers can be improved.
		However, the income statements should not be used for these improvements. The Population Register should be improved with data from the Tax Board; and the Business Register should be improved with the monthly reports from employers that are currently used for the Quarterly Gross Pay survey and are available much earlier.

Using the Income Statement Register, potentially important quality flaws in both the Population Register and the Business Register could be found. Both base registers suffer from undercoverage, and the Business Register also suffers from overcoverage.

D. Integrate the Source with Surveys with Similar Variables

The Income Statement Register has been integrated with the following surveys:

- The *Labour Force Survey*, LFS – employment as employed can be compared with employment indicated by the Income Statement Register. Comparing employment by institutional sector and industry is also important.

- The *Quarterly Gross Pay Survey*, QGP – based on monthly administrative data with gross wages and salaries from all enterprises registered as active employers. It can be used to compare annual gross pay by sector and industry.
- The *Structural Business Statistics Survey*, SBS, contains aggregate wages that can be compared with similar information in the Income Statement Register.
- The establishments according to the Income Statement Register can be compared with the establishments in the Business Register.

The Income Statement Register can be directly linked with the Population Register using the matching key PIN. The Income Statement Register must first be aggregated by employers before it can be linked to the Business Register. This gives the Annual Gross Pay Register, AGP, with gross pay data for enterprises.

Chart 5.16 Information from integrating the source with related surveys – accuracy

Indicator	Quality factor	Description
D1	Is the source good or bad?	The population, units and variables in the Income Statement Register were found to be without quality flaws when compared with the LFS, the QGP and SBS surveys – except that black work is not covered in administrative sources such as income statements.
D2	Are the related surveys good or bad?	Many errors were found in the LFS, the QGP survey and the SBS survey after comparison with the Income Statement Register. Coverage errors in the LFS and SBS were found. Different enterprise units are used in different surveys and in surveys from different periods.
		The sector and ISIC variables were not consistent between different surveys. The coding system for sector and economic activity, ISIC, used in the LFS should be improved. The method for adjusting for black work in the National Accounts should be evaluated.
D3	Can the source improve other surveys?	Many potential problems were found thanks to the income statements in the LFS and QGP surveys. However, income statements and their aggregated version AGP arrive too late to be used as a source for improvement of these surveys. But the AGP can be used to improve the SBS. The quality of the SBS survey can be improved by selective editing and imputation models.
D4	Can the source be combined with other sources?	The Income Statement Register is used for creating some of the variables in the Income Register. Income statements must be combined with other sources to give a full picture of disposable income.
		Income statements alone do not give a complete picture of the economically active population. However, if they are combined with yearly income declarations for enterprises, coverage of employed and self-employed persons is possible. This combination is the basis for the Employment Register.

We tested the work process and these quality indicators on one administrative source, the income statements. Our main conclusions are described in Chart 5.16. The Income Statement Register has been compared with the Labour Force Survey; and the aggregated version of the income statements, the Annual Gross Pay Survey, has been compared with the Quarterly Gross Pay Survey and the Structural Business Statistics Survey. We have found many potential problems and inconsistencies within this system of surveys. These surveys are currently not coherent due to these inconsistencies.

The work with quality assessment is intimately related to the design or improvements of the surveys in the system. We have found the causes of problems and inconsistencies, and the next step should be to reduce the effects of these problems to improve coherence. *Survey system design* refers to the simultaneous work of improving or redesigning a system of surveys. The system-oriented work with quality assessment used here should be the first step in such work with survey system design.

The systems approach has proved to be important. Potential problems in a statistical production system can be detected when we compare many sources and surveys. This is illustrated by the results presented above. The traditional way of working is to consider a survey or administrative source individually. This tradition must be abandoned for quality and efficiency reasons, and a statistical systems approach must be adopted as the general method for working with administrative data.

Conclusions

In the work with quality assessment of the Income Statement Register, we have used methods for the *editing of register data* that are discussed in Chapter 6.

After analysing the indicators A1–D4, we conclude that the input data quality of the income statements is very high. The production process quality is also high, and indicators C1–D4 provide information on the quality of the production system. Coverage errors and lack of coherence are errors that can be measured, but a better strategy is to use the information and improve the system so that these errors are reduced. The errors in the improved estimates will be smaller, but unknown.

This explains why error measures are rare for estimates from register surveys. In contrast to random errors, non-random errors can be measured, and the estimates can thereafter be corrected. But once we have made corrections, we no longer have any quality measure:

- Search for errors with quality indicators A1–D4 (and also E1– E4) above and find the reasons for the errors that have been found.

- Redesign the surveys that have these errors. Calculate new estimates and describe the errors of the old estimators by taking differences between old and new estimates.

- Be satisfied with the fact that the new estimates are the best possible. If there are no other sources or surveys for comparison, the quality cannot be described.

CHAPTER 6

Building the System – Editing Register Data

The editing work for register surveys is different from the current praxis for sample surveys. When sources are combined, *consistency editing* becomes a new task that is unique to register surveys. Errors can be found through consistency editing by comparing data from different sources. Inconsistencies regarding population, statistical units and variables can be found. This chapter presents a number of cases that illustrate methods that can be used. The cases show the importance of consistency editing for the quality of the final statistical register.

For sample surveys, the calculation of sampling errors is a well-known and established method for analysing one important error source. Sampling errors and the methods aiming to reduce them are also central in the work with the design of sample surveys. For register surveys, the systematic work of *comparing different sources* is the method that should be used for analysing quality. Not only will register surveys benefit from such methods, but sample surveys will also benefit as new errors become obvious when sample surveys are compared with registers.

Editing in sample surveys

A sample survey has one main use, and only a limited number of tables are produced because the sampling errors will not permit detailed tabulation. The editing of data can be reduced to prevent 'over-editing'. Errors that do not significantly affect the final estimates can be overlooked.

The main editing phase for sample surveys involves editing of the collected data. Sometimes, contacting the data providers is possible to correct unreasonable variable values. If the editing requires a large amount of resources, this can indicate that the questionnaire needs to be redesigned. Note that errors or suspected errors are always interpreted as errors with respect to variable values. The aim is to replace incorrect or unreasonable values with corrected or reasonable, imputed values.

Register-based Statistics: Registers and the National Statistical System, Third Edition. Anders Wallgren and Britt Wallgren.
© 2022 John Wiley & Sons Ltd. Published 2022 by John Wiley & Sons Ltd.

The scientific literature as a rule discusses only editing of data from surveys with their own data collection – sample surveys and censuses. An overview can be found in Hoogland et al. (2011) and De Wahl (2009). Over-editing is discussed by Granquist and Kovar (1997).

6.1   Editing in register surveys

Two kinds of editing should be done when creating a new statistical register. Chart 6.1 illustrates the work process. First, each source with input data should be edited *alone* when it has been delivered to the NSO. This editing is done in the Throughput Database as described in Section 2.4. The purpose of the editing is to measure and improve the quality of the *variables* in the input data according to the quality indicators in group B in Section 5.4.2.

Different register units will then use this input data for their purposes and combine the new input data with one of the base registers and perhaps other statistical registers. Microdata from different sources can now be compared. With *consistency editing*, the quality of the new register can be measured and improved according to the quality indicators in groups C and D in Sections 5.4.3 and 5.4.4.

Chart 6.1   Editing in sample surveys and register surveys

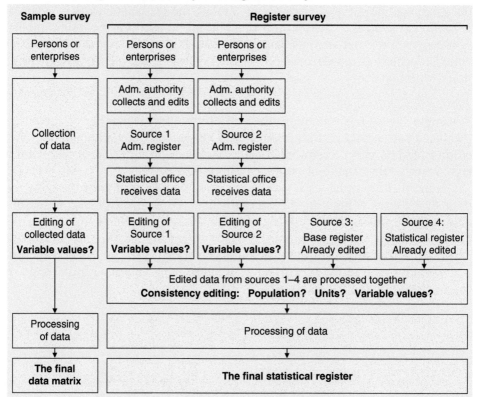

6.2   Editing of a single administrative register

Section 2.4.4 discusses editing of a single administrative register, which continues here with comments on automatic editing and imputation.

A statistical register can be used in many register surveys and a large number of detailed tables can be produced. Thus, it is difficult to define what can be regarded as small errors that can be overlooked. Micro-editing is necessary but editing methods must be developed so that time spent on this work will be reasonable and data quality will be improved.

Values that are rejected during the editing process result in missing values. The missing values should be replaced by imputed values.

In register surveys, the data have been edited first by the administrative authority. Thereafter, every administrative source is edited when the data have been delivered to the statistical office. A source used by many statistical registers should undergo basic editing before it is distributed to the different users within the statistical office. This will save duplication of work, contribute to consistency between registers, and prevent fatal errors. If uncorrected data are distributed to many users, there is a risk that some user will not observe an error.

Software for automatic editing and imputation

Computerised editing routines should be developed to detect and correct errors in extensive registers. The development of methods for editing and correcting microdata in large registers is needed – the methods for editing of sample surveys are not sufficient for register-based statistics.

Pannekoek and de Waal (2005) report that methods have been developed within the EU project EUREDIT for completely automated editing and imputation. Whether these types of program software can be used to edit administrative data remains to be examined. However, automation should never obstruct the development of subject matter expertise.

According to de Waal and Quere (2003), only a small number of countries have developed advanced editing routines that use automatic editing and imputation based on the Fellegi–Holt (1976) paradigm. Such methods could be used to edit administrative sources individually. However, consistency editing of many sources simultaneously gives rise to more methodological problems as errors can be caused both by errors in coverage, in objects and in variables.

Example 1: Automatic editing and imputation. Tax returns for small enterprises. This example illustrates automatic editing and imputation, which is used when the first editing of administrative data is carried out of one source at a time. A total of 464 567 income declarations for a specific calendar year were received from small enterprises using the tax form for small enterprises.

Special software has been developed by Statistics Sweden for this tax form, and the methods used are based on subject matter expertise. The last two variables in the Chart 6.2, 'Deductions for taxation' and 'Taxable income' have the highest quality as they are of great legal and tax administrative importance.

The software starts with these values and then searches for errors in positive or negative signs and summation errors. Finding and correcting most errors becomes rather easy in this register with enterprise data. In the column 'Before editing' in Chart 6.2, almost everything looks absurd. 'Income' is not equal to 'Receipts' minus 'Costs' and not even 'Taxable Income' agrees with 'Income' minus 'Deductions'.

The net errors are shown in the column 'Corrections'. The gross errors in one specific tax return can be more disturbing – there are about 13 000 tax returns that need corrections.

When this source was new to Statistics Sweden, subject matter specialists from different economic surveys looked at the source. One of them said, 'We cannot use this source, we must continue with sending out our own questionnaires.' This illustrates how editing is related to survey design. With no editing or a bad editing method, it is necessary to continue with a sample survey or census. With good automatic editing, a register survey can be carried out instead.

Chart 6.2   Automatic editing and imputation. Tax returns, 464 567 small enterprises

		Before editing SEK billions	After editing SEK billions	Corrections SEK billions	Number of corrections
Receipts	R1	31.017	30.793	-0.224	115
	R2	9.323	9.315	-0.008	23
	R3	0.394	0.392	-0.002	4
Receipts, total		**40.734**	**40.500**	**-0.234**	
Costs	C1	-9.758	-9.746	0.012	22
	C2	-10.363	-0.961	9.402	3
	C3	-6.871	-3.110	3.761	13
	C4	-628.046	-3.027	625.019	7
	C5	-9.979	-9.854	0.125	29
Costs, total		**-665.016**	**-26.696**	**638.320**	
Depreciations etc.	D1	-4.078	-4.097	-0.019	25
	D2	3.880	3.339	-0.541	4
	D3	-3.217	-3.216	0.001	3
	D4	0.537	0.699	0.162	12 857
Depr. etc, total		**-2.877**	**-3.275**	**-0.398**	
Income		**-8.825**	**10.529**	**19.354**	
Deductions for taxation		**-17.789**	**-17.628**	**0.161**	
Taxable Income		**-7.053**	**-7.099**	**-0.046**	

6.3   Consistency editing

There is another, more refined phase in the editing process where data from many sources are edited together. Consistency editing can find additional errors and inconsistencies. Consistency editing is an important phase that is missing in sample surveys with their own data collection. In consistency editing, we are looking not only for errors in variables but also errors regarding population and the statistical units. The errors can be in any of the sources involved.

6.3.1 Consistency editing – is the population correct?

The work of consistency editing consists of analysing a set of data that is a combination of data from different sources. As a rule, different sources have differences in coverage. Because an administrative register should be combined with the base register with the same object type, the case where an administrative source and the base register in question are compared is discussed first. The coverage of different registers is compared and a search for missing and irrelevant categories is undertaken. These issues should be analysed to answer the question: *Is the population correct?*

Coverage errors or matching errors?

Note here that when we match two registers with identity numbers of good quality, the units that did not match can be interpreted as differences in *coverage*. If two registers are matched with a matching key of medium quality, the units that did not match are often interpreted as *matching problems*. However, this can actually be a combination of coverage errors and matching errors.

The example from Chart 5.9, reproduced here, compares a population register for the Galapagos Islands with a test census. The PIN numbers in the test census were collected during the census interviews. If we can conclude that these PIN numbers and the PIN numbers in the Population Register have good quality, then we can say that we have coverage problems in the register. If these identity numbers cannot be trusted, then we do not know if the non-match depends on matching problems or on coverage problems. In this case, there is evidence in Chart 5.9 that the identity numbers have good quality – using names as another matching key gives almost no additional matches. The non-matches are thus generated by coverage problems.

From Chart 5.9   Coverage errors in the population register for Galapagos

Test census 2015 for Galapagos		Population Register for Galapagos 2015	
Population size, N(census)	25 100	Population size, N(register)	30 373
Deterministic matching with register Matching key: Identity number	17 830	Deterministic matching with census Matching key: Identity number	17 830
Probabilistic matching with register Matching key: Names, 85% similarity	67	Probabilistic matching with census Matching key: Names, 85% similarity	67
Undercoverage in the register: persons in the census, but not in the register	7 203	Overcoverage in the register: persons in the register, but not in the census	12 476

Comparing an administrative source with the base register

The quality indicators C1–C5 in Chart 5.7 are relevant when combining data from an administrative register and the relevant base register – the Population Register or the Business Register.

– C1 and C2: Undercoverage or Overcoverage in the base register?

– C3 and C4: Undercoverage or Overcoverage in the administrative register?

– C5: Can the administrative register improve the base register?

Example 2: Comparing the Annual Pay Register and the Business Register.
Aggregate wages and salaries by economic activity and institutional sector are important inputs for yearly and quarterly National Accounts. One quarterly and one annual register survey produce these estimates, and we use these registers to illustrate the work with consistency editing. The administrative Annual Pay Register is generated by a system where all employers report during January after the income year. This source is regarded as a source with good quality. An administrative register of good quality can be used almost as it is for statistical purposes. If content and coverage are sufficient, using the administrative register and producing statistics of good quality will be relatively easy.

The different steps of the yearly survey are illustrated in Charts 6.3a–6.3c. The Annual Pay Register is compared with the register with legal units coded as active employers in the Business Register.

Chart 6.3a   The Business Register and the Annual Pay Register

1. Business Register			2. Administrative Annual Pay Register				Sector:
BIN	Sector	ISIC	BIN	WagesYear	Source	Prel-tax	
BIN02	6	52	BIN01	25	I	8	1 Non-financial enterprises
BIN03	1	51	BIN03	1 667	I	544	2 Financial enterprises
BIN04	7	91	BIN04	796	I	252	3 Government
BIN05	1	70	BIN05	2	P	689	4 Municipalities
BIN06	1	45	BIN06	92	I	29	6 Self-employed
BIN07	1	51	BIN07	4 758	I	1 565	7 Non-profit organisations
BIN08	1	60	BIN08	39	P	12	
BIN09	1	28	BIN09	452	I	142	Source
BIN10	1	74	BIN11	289	P	95	I Internet
BIN11	1	27	...				P Paper form
...			**Count**	**305 411**			
Count	**331 518**						

The administrative Annual Pay Register is first edited alone to find incorrect or unreasonable values of aggregated wages and salaries (WagesYear). Preliminary tax (Prel-tax) should be between 30% and 35% of WagesYear, and this relation is used to edit WagesYear. For the enterprise with identity number BIN05, we find a 1000-factor error; 2 is therefore replaced with 2 000. The registers in Chart 6.3a are then matched with the identity numbers BIN. Observations with the same value of BIN are combined in a new register that is shown in Chart 6.3b. The variable W-imp is created to show imputed values of WagesYear in Chart 6.3b.

Chart 6.3b   After matching the Business Register and the Annual Pay Register

(1)	(2)	(3)	(4)	(5)	(6)
BIN	Sector	ISIC	BIN	WagesYear	W-imp
*	*	*	BIN01	25	0
BIN02	6	52	*	*	*
BIN03	1	51	BIN03	1 667	0
BIN04	7	91	BIN04	796	0
BIN05	1	28	BIN05	2 000	1
BIN06	1	45	BIN06	92	0
BIN07	1	51	BIN07	4 758	0
BIN08	1	60	BIN08	39	0
BIN09	1	28	BIN09	452	0
BIN10	1	74	*	*	*
BIN11	1	27	BIN11	289	0
...					

	Name	Count	Missing
(1)	BIN	365 061	33 543
(2)	Sector	365 061	33 543
(3)	ISIC	365 061	33 543
(4)	BIN	365 061	59 650
(5)	AggrWages	365 061	59 650
(6)	W-imp Imputed values, wages		

A large amount of *non-match* is found when the 331 518 observations with active employers in the Business Register are matched with the 305 411 observations in the administrative Annual Pay Register. There are 33 543 observations in the administrative register that are missing in the Business Register; this is an indication of undercoverage according to quality indicator C1.

We also find that 59 650 legal units that were included in the Business Register as 'active' employers did not exist in the Annual Pay Register. This is a measure of overcoverage according to indicator C2. The conclusion is that the Annual Pay Register can be used to improve the Business Register (indicator C5). The non-match is shown to the right in Chart 6.3b. The non-match above gives rise to *missing values* (*) in the variables Sector and Economic Activity (ISIC). After imputations, the final statistical Annual Pay Register in Chart 6.3c can be created.

The non-match in Chart 6.3b arises because the Business Register and the administrative Annual Pay Register give conflicting pictures of the population of active employers. Such conflicts always arise when different sources are compared, and a decision must be made regarding which source should be trusted.

If the Business Register in columns (1)–(3) in Chart 6.3b is trusted, then enterprise BIN01 should be excluded and missing values should be imputed for enterprises BIN02 and BIN10.

If instead the Annual Pay Register in columns (4)–(6) in Chart 6.3b is trusted, then enterprises BIN02 and BIN10 should be excluded and missing values for enterprise BIN01 should be imputed as in Chart 6.3c.

Chart 6.3c   The statistical Annual Pay Register

BIN	Sector	S-imp	ISIC	I-imp	WagesYear	W-imp
BIN01	6	1	01	1	25	0
BIN03	1	0	51	0	1 667	0
BIN04	7	0	91	0	796	0
BIN05	1	0	28	0	2 000	1
BIN06	1	0	45	0	92	0
BIN07	1	0	51	0	4 758	0
BIN08	1	0	60	0	39	0
BIN09	1	0	28	0	452	0
BIN11	1	0	27	0	289	0
...						

Count   305 411

The choice between these alternatives should be based on a clear understanding of *the administrative system*. In this case, there is no cost involved to remain as 'active employer' in the system and report and pay 0 every month.

Thus, enterprises that have been active employers in the past may choose to remain as 'active' even if they have no employees.

For this reason, we have chosen to trust the Annual Pay Register.

A system-oriented approach

Chart 1.15 lists the four principles that describe how administrative registers should be used for statistical purposes. In the example above, the first line of the transformation principle is illustrated. We now turn to the second line. Why should we use many sources and how?

---

1. Transformation principle
These administrative registers should be transformed into statistical registers. All relevant sources should be used and combined during this transformation.

Let us review the Annual Pay survey above. A register with aggregate wages for all enterprises was combined with the Business Register to create a statistical register with sector, ISIC, and aggregate wages. Using this register, estimates of aggregate wages by sector and ISIC are derived and delivered to the yearly National Accounts. The relevance and accuracy of these estimates are regarded as very high.

During this work, we focused our thinking only on the Annual Pay survey – this is an example of the common *one survey at a time* thinking. A system-oriented approach would also include the following two aspects:

– Can we use more parts of the production system to improve the Annual Pay survey? This is quality indicator D4 in Section 5.4.4.

– Can we use the Annual Pay Register to improve the production system? This is quality indicator D3 in Section 5.4.4.

To answer these questions, the statistical Annual Pay Register is combined with the Quarterly Pay Register that has similar variables. This register is based on monthly tax reports from all employers on aggregated wages and salaries. These reports are delivered two weeks after the income month. When the Annual Pay Register is created, the Quarterly Register is available and can be used.

We now have a system of three statistical registers. If these three registers are combined, we obtain the register shown in Chart 6.4a.

Chart 6.4a   Combining the Business, Annual and Quarterly Pay Registers

Columns (1)–(3) from Business Register, (4)–(5) from Annual Register and (6)–(10) from Quarterly Register

(1)	(2)	(3)	(4)	(5)	(6)	(7)	(8)	(9)	(10)
BIN	Sector	ISIC	WagesYear	W-imp	WagesQ1	WagesQ2	WagesQ3	WagesQ4	SumQ1-Q4
BIN01	*	*	25	0	25	0	0	0	25
BIN02	6	52	*	*	*	*	*	*	*
BIN03	1	51	1 667	0	300	371	384	610	1 665
BIN04	7	91	796	0	233	248	184	130	795
BIN05	1	28	2 000	1	407	403	412	852	2 074
BIN06	1	45	92	0	0	10	32	49	91
BIN07	1	51	4 758	0	1 093	1 236	1 684	1 214	5 227
BIN08	1	60	39	0	35	0	0	4	39
BIN09	1	28	452	0	99	120	119	112	450
BIN10	1	74	*	*	*	*	*	*	*
BIN11	1	27	289	0	65	65	65	93	288
...									
BIN20	2	65	627	0	43 451	41 407	43 964	39 442	168 264
...									
BIN30	1	74	2 675	0	361	461	606	639	2 067
BIN40	1	73	0	0	349	256	0	0	605
...									

Count:	34 574 = undercoverage in Business Register	366 092 =
	57 286 = overcoverage in the Business Register	= total number of observations

When we compare the first 11 rows of columns (4) and (10) in Chart 6.4a, we find that aggregate wages are almost the same with the exception of enterprise BIN07. We have added three observations that show large differences. For enterprise BIN20, we found that a number of insurance companies (ISIC = 65) report insurance benefits in the same way as wages in the monthly tax reports, as tax is paid in the same way as for wages. In the yearly taxation system, insurance payments and wage payments are handled separately; but these two payments are combined in the monthly system. The staff at Statistics Sweden was not aware of this until the Annual and Quarterly Pay Registers were combined.

In addition, the differences are large for enterprises BIN30 and BIN40. This is an example of a mid-year takeover – after the takeover enterprise BIN30 sends in tax reports for all employees at both units. As a result, the Annual and Quarterly surveys will give inconsistent estimates of aggregate wages by industry.

Hence, the answer to the first question is: Yes, the Annual survey can be improved if we also use the Quarterly survey. The imputed value for enterprise BIN05 can be replaced with the value from the Quarterly survey; the values from the Quarterly survey give better estimates by industry for the last two enterprises in Chart 6.4b and coverage has been improved with 2 812 more enterprises.

Chart 6.4b   Final version of the statistical Annual Pay Register

Aggregate wages are based on the best of Annual and Quarterly sources

BIN	Sector	S-imp	ISIC	I-imp	AggrWages	W-imp	W-source
BIN01	6	1	01	1	25	0	Y
BIN03	1	0	51	0	1 667	0	Y
BIN04	7	0	91	0	796	0	Y
BIN05	1	0	28	0	2 074	0	Q
BIN06	1	0	45	0	92	0	Y
BIN07	1	0	51	0	4 758	0	Y
BIN08	1	0	60	0	39	0	Y
BIN09	1	0	28	0	452	0	Y
BIN11	1	0	27	0	289	0	Y
...							
BIN20	2	0	65	0	627	0	Y
...							
BIN30	1	0	74	0	2 067	0	Q
BIN40	1	0	73	0	605	0	Q
...							

S-imp =	Sector imputed	6 795	only in Annual
I-imp =	ISIC imputed	299 199	both in Quarterly and Annual
W-imp =	AggrWages imputed	2 812	only in Quarterly
W-source =	Source for AggrWages	308 806	total number of observations

We turn to the second question: can the Annual survey be used to improve some parts of the production system at the NSO?

During work with the Annual Register, we found coverage errors in the Business Register. Since the Business Register is used as a basis for all business statistics, the accuracy of many economic surveys will be improved if data from the Annual and Quarterly surveys can reduce these coverage errors. When data from the Annual and Quarterly registers were compared, we found a serious misunderstanding regarding insurance companies. Both wages and insurance benefits were reported as wages in the monthly tax reports on aggregate wages. With more variables for the same enterprise, we have greater possibilities of finding and correcting errors in variables. We can also obtain information on units in the population that have changed in some important way, e.g. that one enterprise has taken over another.

During the work with register-statistical processing, we have had the opportunity to observe errors and anomalies in the data we use. These observations should be developed into systematic work with quality assessment.

The examples regarding the Quarterly and Annual Pay Registers illustrate many possible ways of using an administrative source. We should not only consider one source and one register survey at a time, but also consider other sources that could be combined together and whether a source can be used for other purposes.

Administrative source 2 in Chart 6.5 is the main source for the Quarterly Pay Register, but it can also be used for the Annual Pay Register. When we use the Quarterly register for quality assessment, we find coverage problems with the Business Register. When we use the Annual Pay register for quality assessment, we find errors regarding data from insurance companies in source 2. Errors can be reduced if sources 1, 2 and 3 are combined and used in new ways. The example illustrates that a system-oriented approach is essential when designing register surveys.

Chart 6.5   All relevant sources should be considered simultaneously

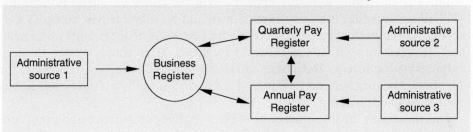

Searching for missing categories in one source

The first check is made when the administrative register is delivered from the administrative authority to the statistical office. How many observations were sent? How many were received? Did we obtain all variables? Do the observations have the proper format? This is how the completeness of the delivery is checked.

In many cases, the administrative authority that delivers the register to the statistical office has received the data from a number of regional offices. The statistical office should check if all regions are represented in the delivery.

Macro-editing, where the new register is compared with the old version, is a method that should be used to check the completeness of the new data that have been delivered.

Searching for irrelevant categories

Administrative systems are designed for administration, not for statistical purposes. Consequently, the administrative object set may contain categories that should not be included in the statistical register. How should this problem be handled by the statistical office? An example highlights some aspects of general interest.

Example 3: Norwegian hospitals, do they have activities in Sweden?
Norwegian hospitals started to encourage Swedish nurses to come to Norway and obtain well-paid jobs at Norwegian hospitals. However, the Swedish

nurses planned to return to Sweden after a period in Norway. Since they want to receive social benefits from Sweden, they want to pay Swedish taxes even when they work in Norway. Approximately 20 Norwegian hospitals have therefore registered legal units in Sweden. These legal units are used only for paying wages to the Swedish nurses and they send their preliminary tax payments to the Swedish Tax Agency. This administrative practice has the following consequences:

- About 20 legal units in the Business Register managed by the Swedish Tax Agency have no production in Sweden.
- About 1 500 jobs in the Swedish Income Statement Register concern jobs in Norway. The total yearly wage sum is a few hundred million SEK.
- More than 1 000 Swedish nurses are registered as living and working in Sweden, but they actually live and work in Norway.

This is typical for an administrative source; a new, complicated kind of administrative transactions suddenly appears. Will the statistical office notice this? Will all surveys that use this source notice and handle this new category correctly? At the perfect statistical office, the new kind of legal units is immediately noticed by staff at the Business Register. After contacts with the unit responsible for foreign businesses at the Swedish Tax Agency, experts at the National Accounts are contacted and a decision is made on how each survey should handle the new category.

This illustrates the importance of subject-matter competence and good contacts with the administrative authority. It also shows the importance of teamwork among staff working with different surveys.

Consistency editing, where many sources are compared, can be used to find irrelevant categories. When enterprises are found that make large payments to employees but have no turnover, this category can be examined, and the Norwegian hospitals will be found.

Measuring overcoverage through register activities

If an enterprise has not delivered any tax report or made any tax or social insurance payments during a long period, the enterprise can be coded as 'inactive' in the statistical Business Register and can be treated as overcoverage. We recommend that the activity during each calendar year is checked for all enterprises in the yearly Business Register.

Similarly, if a person has no signs of activity in any relevant administrative register during a long period, the person probably has emigrated and can also be regarded as overcoverage. Statistics Sweden measures overcoverage in this way, and the overcoverage in the Population Register for 2014 was estimated to be about 68 000 persons.

6.3.2   Consistency editing – are the units correct?

Errors in units refers to the belief that we are comparing data concerning the *same* unit in different sources. However, the data we compare come from *different* units which have the same identity number. This will be the case when we have false positive matches, or when we have created derived units incorrectly. When we might have errors in units, we should not correct or impute variable values until we have checked that the units are the same.

The unit problem is a serious problem in economic statistics. Enterprises merge and split and the old identity number BIN is used for the new changed units. When such changes happen during the year, some monthly values refer to the old unit and some to the new, and the values will not be comparable over time. The editing method to find such problems consists of following monthly or quarterly time series and searching for level shifts. Yearly data can also be compared with sums of monthly or quarterly values.

Example 4: Editing legal units with different wage sums
Two sources with gross annual pay are integrated in this example. One source is the Annual Pay Register (AGP) and the second is the Quarterly Pay Register (QGP) where twelve monthly values have been summed into gross annual pay. In Chart 6.6 the inconsistencies between these sources are illustrated.

Chart 6.6   Comparing gross annual pay in a quarterly and yearly source, microdata

BIN	ISIC	Gross annual pay,	SEK million	
		QGP	AGP	QGP–AGP
BIN5	41	259	663	–404
BIN6	43	115	0	115
BIN7	43	112	0	112
BIN8	42	175	0	175
BIN9	29	84	110	–26
BIN10	25	25	0	25
BIN11	47	681	731	–50
BIN12	46	50	0	50

These observations show errors in units; the four observations BIN5–8 have merged into BIN5 and BIN9–10 and BIN11–12 have also merged. Each observation has the same ISIC when gross pay in the two surveys is compared. But because the legal units have changed, estimates by ISIC will be different.

Example 5: Editing legal units with different turnover
Another problem is that many enterprises, each with their own BIN identity, belong to *a group of legal units*. One legal unit within the group can be used for VAT reporting and payments for the enterprise group, while another legal unit is perhaps used for social security reporting and payments for the whole group, etc.

Chart 6.7 has taken data regarding the first five legal units in Chart 3.3, where turnover has been reported in three different sources. These five legal units are large corporations with competent and reliable staff. Each source has been edited individually and errors due to positive or negative signs, summation errors, scanning errors, etc. have been eliminated. So, why are the turnover values for each legal unit so extremely different?

The five values from the SBS survey are all incorrect. The persons at the large corporations have misunderstood the unit or group of units about which they should give information. The Yearly Income Tax return (YIT) is the basis for the Structural Business Statistics survey (SBS); then the variable that should be measured consists of the values in the YIT column. The company with BIN = 160002 reports turnover under a different BIN number in the YIT data set. The companies with BIN = 160001, 160003 and 160005 report turnovers for different groups of units in the YIT source than in the VAT source. All values in the two administrative sources are probably correct but refer to different groups of units.

Chart 6.7   The unit problem in business data
Yearly turnover according to three sources

BIN	SBS	YIT	VAT
160001	7 179	11 941	8 089
160002	2 954	0	0
160003	843	3 561	918
160004	5 514	2 888	2 895
160005	26	538	2 536
...	...	...	...

The method that should be used here is to integrate sources with similar variables and then compare these variables to find probable errors in units.

These erroneous units are thereafter 'edited' and replaced with *derived units* consisting of groups of legal units.

6.3.3   Consistency editing – are the variables correct?

In this example, the Quarterly Pay Register (QGP) and the Annual Pay Register (AGP) are compared again in Chart 6.8. The method of finding errors in variables is the same as before; sources with similar variables are integrated and differences are analysed. But the sources have already been edited; each source was first edited alone in the search for errors in variables. Thereafter, the sources were checked if there are errors in the units. This means that we are now looking for errors in variables that have other causes.

Chart 6.8   Comparing gross annual pay in QGP and AGP, microdata

BIN	ISIC	Gross pay, SEK million		
		QGP	AGP	QGP – AGP
BIN1	65	5 956	265	5 692
BIN2	65	1 455	310	1 145
BIN3	65	817	1	816
BIN4	65	328	8	320

These observations are examples of errors in variables. After checking, gross pay in the quarterly source was discovered to contain both wages and insurance benefits (ISIC 65 = Insurance).

The conclusion is that for ISIC 65, the quarterly source suffers from a serious relevance error. Chart 6.9 shows the error in the estimated gross pay.

Chart 6.9 Comparing gross annual pay in QGP and AGP, macrodata

| ISIC | Gross pay, SEK million | | | Number of |
	QGP	AGP	QGP – AGP	enterprises
65	16 113	8 469	7 644	567
...	...	...	...	...
All	1 246 593	1 241 138	5 454	307 230

The main part of the difference in total gross pay between the two surveys is due to the problem with ISIC 65.

The error is substantial in the estimate of gross pay in the insurance industry (ISIC = 65). This error was discovered for the first time when we suggested that the QGP and AGP surveys should be compared at the micro level.

6.4 Case studies – editing register data

Two case studies are presented below that illustrate the editing work of different registers at Statistics Sweden.

6.4.1 Editing work within the Income and Taxation Register

The Income and Taxation Register (I&T) is the first example of how editing work can be organised. This register is used to describe the distribution of income and taxation for individuals and families based on tax self-assessments and other administrative registers.

The register's variables are also used in the FASIT micro-simulation model. This simulation model is used by the Ministry of Finance to study the effects of planned changes in taxation and transfer payments. The register must fulfil many important quality requirements at a micro level, such as the income and taxation values for individual persons and families which must be complete and consistent. Extensive work with editing and correcting variable values is necessary to ensure that there are no strange simulation results.

A description of the work to create the register is given in Section 1.4.4, where the I&T Register is used to illustrate how administrative registers are transformed into statistical registers. Chart 1.12 lists the different source registers used.

Step 1 – Editing of each administrative register
In total, around 30 administrative registers are received every year. The first step is to edit every one of these as follows.

Firstly, the *variable descriptions* that come with the new administrative registers are checked. These variable descriptions generally change every year; variable names may have changed and new variables may have been added. The variable description of complicated variables may be misunderstood, which will generate errors. Therefore, maintaining close contact is important with the persons at the administrative authority who can give the necessary explanations.

Thereafter, the *extreme values* of the quantitative variables are studied. These are compared with the *previous year's values* at both an aggregated and individual level. Some variables, such as sickness benefit, have a ceiling value that can be used for these checks. Next, *logical checks* are carried out to verify that the totals given are actually the sums of their parts.

Certain variables are reported from subordinate authorities to the central authority, which in turn delivers the data to Statistics Sweden. Checking that all the subordinate authorities have provided data is appropriate in such situations. For example, data on social assistance is usually missing for a few municipalities every year. Those municipalities that have not provided data must be documented in the I&T Register; and imputations can be carried out for these missing values, usually using the previous year's social assistance.

Editing work affects not only register quality. If the work is organised so that several persons share responsibility, the editing can contribute to cooperation and the exchange of experience within the team. This will increase *subject-matter expertise* and indirectly the quality of the register. Subject-matter expertise is strengthened when documenting the work and taking measures to correct data.

Contacts with suppliers have several important effects. Firstly, the staff at the administrative authority should be informed about how and for what purpose their data are used at Statistics Sweden. The staff at the authority should understand the consequences of lack of quality for users of the statistics. Contacts with the suppliers are also important for the subject-matter expertise of the staff at Statistics Sweden. This is why the staff working with the I&T Register have regular meetings twice a year with the Swedish Tax Agency. These contacts are also used in the important work of identifying new administrative sources.

Step 2 – Final checking of the entire register
In the first step, all the data from each authority are checked. In the next step, all variables from all sources are combined in one total register so that the different sources can be compared through *consistency editing*. All the derived variables are then formed. In this way, new consistency checks can be carried out, i.e. the sum of all variable values from different sources agrees with the sum from another source. Additional errors can also be identified. The total register consists of around 9 million observations with 500 variables.

Example: A subset of four variables was checked by macro-editing with respect to sums and number of persons with values for these variables. Comparisons were made with corresponding variables from the previous year. Everything looked quite normal. A derived variable was created with variables in this subset describing a special kind of income:

$$Income = Variable_1 + Variable_2 - Variable_3 - Variable_4$$

About 120 000 persons were found to have a negative value regarding this kind of income. In the previous year, only approximately 1 600 persons had negative income. After checking, $Variable_3$ and $Variable_4$ were found to be sums of monthly values, where the value for April had been counted twice by the administrative authority.

An important lesson from this example is that the error was found through the derived variable. Another lesson is that the work done by an administrative authority when preparing a delivery to the statistical office can generate errors. Close cooperation is necessary to reduce this error source.

Step 3 – Checking estimates
In this step, the whole register is used as the basis for forming all important tables. Estimates are checked and compared with the previous year's values. In addition, a number of simulations are carried out using the FASIT micro-simulation model for the sole purpose of testing data quality. If, for example, the housing benefits remain unchanged in the model, then the model should generate model values that agree with the previously produced tables.

6.4.2   Editing work within the Income Statement Register
The register of all income statements for social security payments is used to calculate region-specific and industry-specific wage sums. It is used when the Activity Register and the Employment Register are created. This section gives an account of the editing work carried out on the definitive income statements, which are received by Statistics Sweden up to October. The income statements are checked by those responsible for the Income Statement Register. The edited register is then used as the source for other registers within the register system.

Checking population definitions
The first step in the editing process involves checking that the number of received income statements agrees with those sent from the Swedish Tax Agency.

The second step is to create a data matrix with the final income statements according to all the amendments in the consignment. The Swedish Tax Agency does not change input data. When the data provider (in this case the employer) submits amendments to the agency, new observations are created equivalent to deletion, amendment, or replacement of previous observations. Processing is therefore required in the register to remove invalid observations and to check for duplicate observations.

The third step in the editing work is to check all identities. As income statements contain individual and enterprise identities, both the personal identification numbers PIN and organisation numbers BIN should be checked. Around 7 600 personal identification numbers were incorrect, of which 5 000 could be corrected automatically.

The fourth step involves matching the personal identification numbers in the income statements with those in the Population Register, and matching the enterprise identities against the Business Register. In both cases, several non-matching observations are found. The Income Statement Register contains personal identification numbers that are missing in the Population Register as well as enterprise identities that are missing in the Business Register.

Checking variable values
In the fifth step, deviation errors are checked using 16 different probability checks. The relation between earned income and tax is used in several ways. In

addition, a search is made for observations with extremely high earned income or tax. Around 5 000 observations with extreme values are detected from these checks. These are checked in a simple way and only a few are checked with the Swedish Tax Agency. After these checking stages, each income statement is accepted, replaced by a new statement, or taken out of the register.

Checking the most important variable

The most important phase in editing work involves checking that employed persons are linked to the correct establishments. This link is crucial for the whole register system because it enables reporting gainfully employed persons by industry sector and region. Difficulties arise with this link when enterprises have more than one establishment. Although the employer has a duty to indicate the establishment on every income statement, this information is often missing and sometimes implausible. Implausible establishment numbers are identified by comparing the number of employees with corresponding data in the Business Register and with data from the previous year's version of the Income Statement Register. Plausibility in terms of commuting distance is also considered.

When an establishment is missing or appears implausible on the income statement from enterprises with more than one establishment, the employer is contacted via a special data collection using a register update questionnaire. Those responsible for the Income Statement Register work together with those responsible for the Business Register to capture changes regarding the establishment's municipality code and industrial classification code.

Output editing

The Income Statement Register is used as a source for the Employment Register. By checking the output from the Employment Register, the quality of the Income Statement Register is also checked. Detailed tables with employed persons by industry and municipality are assembled and compared with the previous year's tables. Deviations are checked and the results of these checks are documented. This documentation is very useful, since many users make inquiries questioning the results following publication. Where documentation exists, those who are in contact with the users can respond that 'we have checked, and the results are correct as far as we can see'.

6.4.3   What more can be learned from these examples?

The examples above show that the administrative data received at Statistics Sweden may contain errors that require checking at the micro level. Once these errors have been detected, they are often easy to correct. The requirements of the checking procedure depend on how the register is used. Statistics Sweden's statistical registers are often used for research. The quality at a micro level needs to be higher for such advanced analytical needs than when only simple tables are produced; and higher demands are made on the checks. High requirements are primarily made with regard to longitudinal studies.

Subject-matter expertise and contacts with suppliers
An overall conclusion is that subject-matter expertise is of great importance for the effectiveness of the editing and checks. For surveys with their own data collection, it is sufficient to be familiar with the survey in question, which is rarely changed. With register surveys, however, familiarity with the administrative system that generated the data is necessary. An administrative system can contain many complicated variables that change often.

The example also shows the importance of cooperation and development of expertise within the working group that receives the administrative registers, as well as maintaining good contacts with the authorities supplying the data. Furthermore, cooperation between different teams working with related registers should be encouraged so that the administrative data are used effectively.

If the staff at the statistical office 'live with the data at micro level', the learning process is ongoing. This leads to better subject-matter expertise. This learning process is strengthened by close contacts with users.

Additional data collection may be necessary
When a variable in the administrative data appears to have insufficient quality for statistical purposes, additional data collection may be necessary to attain a sufficiently high level of quality. One example is the editing work of checking that employed persons are linked to the correct establishments in the Income Statement Register. To achieve sufficient quality, some employers are contacted via a special data collection using a register update questionnaire.

6.5   Editing, quality assessment and survey design

Editing is the systematic work to find obvious and probable errors. Editing is thus important for learning about the quality of each administrative source and the final statistical register. Quality issues are also of central importance in the work with survey design. This means that editing, quality assessment and survey design are closely related topics.

6.5.1   Survey design in a register-based production system
There are two approaches to survey design when an NSO gains access to microdata from administrative registers:

- With the *traditional approach*, we start with the *survey content* we want. For example, we want to conduct an income survey and then we start planning for an income register. We search for administrative sources that can be used in creating an income register and develop the register-statistical processing methods that should be used.

- With the *systems approach*, introduced in Laitila, Wallgren and Wallgren (2012), we systematically analyse *each administrative source* and try to find out *how* it should be used within the production system or register system. For example, if we analyse income self-assessment from persons as in

Section 5.7, we will find that this source can be used in many ways. It can be used for an Income Register and for sample surveys regarding income of households. It can also be used to improve coverage of the Population Register, the Job Register and the Business Register. Also here we develop the register-statistical processing methods that should be used, but the scope is wider – we not only develop one register survey, we improve several parts of the production system.

Survey design consists of the efforts to maximise the quality of estimates generated by a specific survey, subject to cost or budget constraints. By quality we generally mean accuracy, but other quality dimensions can be included such as relevance, timeliness, comparability, and coherence. Biemer (2010) uses the term *'fitness of use'* for this broader quality concept.

Redesign of the Annual Pay and Quarterly Pay surveys

With the example in Section 6.3 we will illustrate that editing, quality assessment and survey design are closely related topics. The Quarterly Pay Register for the fourth quarter year t is completed during February year $t + 1$, long before that the Yearly Pay Register for year t is completed during September year $t + 1$. Due to this time difference, no one had thought of that it could be a good idea to use the Quarterly Pay Register for the yearly survey. But we suggested that the Annual and Quarterly Pay registers and the Business Register should be combined, and that consistency editing should be done. After consistency editing the following quality problems were found:

- Coverage problems, mainly in the Business Register but also in the two pay registers (Chart 6.4a and 6.4b).
- Missing values for the variables *Sector* and *ISIC* in the Business Register (Chart 6.4a).
- Relevance error for the variable *wage sum* in the Quarterly Pay Register (Chart 6.8 with Insurance companies).

With knowledge of these quality deficiencies, both the Annual and Quarterly Pay surveys can be redesigned. Through the following register-statistical processing the wage sum surveys, and the Business Register will be improved.

- Enterprises that are active with tax reporting and payments should be included in the Business Register. The methods used for updating the Business Register should be redesigned for this purpose. This is discussed in Section 10.5.
- The coverage of the register population that is used for the Annual Pay Register can be improved with about 3 000 small enterprises from the Quarterly Pay Register.
- There are about 35 000 missing values regarding the variable *Sector* that can be replaced with imputed values based on information in the identity numbers BIN. The Business Register should be responsible for this.
- Define *final wage sum* with the yearly source when possible, otherwise use the monthly source. Most enterprises have both a yearly wage sum based

on monthly wage sums, and a yearly wage sum. As the yearly wage sum is judged to be of the best quality, this source is given priority.

– There are about 45 000 missing values regarding the variable kind of economic activity *ISIC* in the Business Register. The staff working with the Business Register is responsible for handling this problem. Imputed values can be created with random imputations.[1] Different imputation models for each sector can be developed, and imputed values based on a nearest neighbour method can be used where size is measured with wage sums.

6.5.2 Survey design – management problems

We have tested the systems approach to survey design by analysing microdata from five surveys. The intention was to design a new survey where *productivity* by industry in the sector of non-financial enterprises would be estimated using estimates of *value added* from the Structural Business Statistics survey (SBS) and estimates of *hours worked* from the Labour Force Survey (LFS). Analysing the quality of these estimates required analysis of the registers constituting the links between the LFS and the SBS. This means that the Population Register, the Job Register and the Business Register were also analysed (Chart 6.10).

After consistency editing of the sources in Chart 6.10, several kinds of coverage errors and variable errors were found. It is possible to link persons in the LFS with the enterprise in the SBS survey and compare coverage and ISIC-coding.

To find errors is not the problem – the main problem here is how to organise the redesign of the system of registers and surveys in Chart 6.10. There are at least eight managers involved at different levels in the organisation. Five surveys and registers should be redesigned in a very precise and coordinated manner so that productivity can be estimated with the best possible quality. This places high demands on competence and the ability to work together.

Chart 6.10   The system of registers and surveys that was analysed

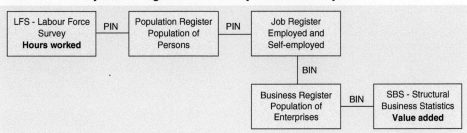

---

[1] In Chapter 12 in the second edition of our book, Wallgren and Wallgren (2014), there is more on missing values in registers.

6.5.3   Total survey error in a register-based system

The total survey error describes all errors that give rise to lack of accuracy. The sampling error is always measured in sample surveys, but the other non-sampling components are seldom measured. However, the non-sampling errors should always be considered during the design process. The total survey error is discussed by Groves and Lyberg (2010) and is considered to be *'the conceptual foundation of the field of survey methodology'*.

Register surveys should also be included in the survey methodology and this area is becoming increasingly important as the use of administrative data increases. What similarities and differences can be found if we compare the sample survey based ideas in Biemer (2010) and Groves and Lyberg (2010) with the example in Chart 6.10 where all surveys are dependent on registers?

The most important difference is that Biemer, Groves and Lyberg discuss *one* (sample) survey at a time; it is *one* survey that should be designed so that the total survey error is minimised under the budget constraints. In the example above with register surveys, a *system* of surveys is considered. A sample survey, the LFS, is included in our system; but some survey error components of the Swedish register-based LFS are determined by undercoverage in the Population Register. So we cannot design the LFS alone. We must simultaneously consider the design of the Population Register and other parts of the Swedish production system that are used together with the LFS.

Another difference is that we can measure many important non-sampling errors of the LFS and other surveys in the system. We can do this by integrating data from different parts in the system. We compare the Population Register and the Job Register and find coverage errors; we compare the Job Register and the Business Register and find more coverage errors. And we can compare classification of economic activity in a number of surveys and describe the inconsistencies in the system. We do not have to use quality indicators only; we can measure relevant quality components directly.

CHAPTER 7

Building the System – The Population Register

The statistical Population Register plays a threefold role in a country with a register-based production system.

- The register is a survey that gives information corresponding to the short form in a traditional census, and persons and households are georeferenced with the register. *Population statistics* for small regions and categories can be produced very timely.
- The register is used to create *frame populations* for sample surveys and *register populations* for register surveys in all the surveys of persons and households.
- The register is the source of several *standardised variables* that are used in sample surveys and register surveys regarding persons and households. These variables are also used as auxiliary variables in sample surveys.

This chapter first explains the methods used when we have a Population Register based on full information; that is, we have all the information needed from administrative sources. We have information on all demographic events together with time references – we know when each event occurred.

Thereafter, this chapter discusses methods that can be used in new register countries that do not yet have population registers based on full information.

How to produce consistent register-based statistics
In a traditional census, interviewers collect data regarding all census variables during the interview. The census data can therefore be used to produce consistent tables with estimates that describe the census variables for small areas and small demographic groups.

In a register-based system, consistent census-like tables should be produced in a simple but efficient way. All populations in the production system should be created with the help of one of the base registers. Consistency regarding populations is achieved if all register surveys that replace the census use the same version of the Population Register for population. This version is used as a *standardised population* and defines the object set for many other registers.

Register-based Statistics: Registers and the National Statistical System, Third Edition. Anders Wallgren and Britt Wallgren.
© 2022 John Wiley & Sons Ltd. Published 2022 by John Wiley & Sons Ltd.

The first step in the process involves the creation of a standardised population of persons and households by the NSO team responsible for the Population Register. The standardised population can, for example, be defined as the resident population on 31 December. The population for 31 December of year t is created during year $t+1$ when all information regarding year t is available. Statistics Sweden creates this population during February each year.

The next step is to use this standardised population as the *register population* in the other statistical registers described in Chart 7.1, when the teams responsible for the other registers create the Employment, Education, Income and Occupation Registers at different points in time and with different sources. In this way, five different register teams coordinate their work and produce five consistent registers during a time span from February until November each year.

Chart 7.1   Decentralised but coordinated process to create registers on persons

Administrative sources 1	⇒	Statistical register 1: **Standardised population**				
		Standardised population	+	Administrative sources 2	⇒	Statistical register 2: **Employment Register**
		Standardised population	+	Administrative sources 3	⇒	Statistical register 3: **Education Register**
		Standardised population	+	Administrative sources 4	⇒	Statistical register 4: **Income & Taxation Register**
		Standardised population	+	Administrative sources 5	⇒	Statistical register 5: **Occupation Register**

Because the administrative sources 2–5 do not overlap in terms of variables, the work to create statistical registers 2–5 can be done independently of each other.

Chart 1.14 contains examples of completely consistent tables produced in this way. The tables describing population, employment, education, income, and occupation for small areas, age classes and genders are completely consistent.

Register population

The concept *register population* refers to the object set in the register that has been created for the register survey in question, that is the population that is *actually* being surveyed. This concept corresponds to the concept *frame population* used for sample surveys.

There are, however, important differences between a frame population and a register population. A frame population is defined *before* the data collection, while a register population is created *after* the reference period when all administrative data for the period have been received.

It is important to distinguish between the concepts *frame population* and *register population*. Frame population is a term that should only be used for sample surveys. Register populations are defined later and can be based on more and better information. Thus, these two kinds of population have different coverage problems – register populations will as a rule have smaller coverage errors.

7.1 Inventory of sources

What information do we need to create a statistical Population Register? Information about *demographic events* such as births, marriages, deaths and migration should be included in both the administrative and statistical Population Registers. Coverage problems, underreporting and delayed reporting of internal and external migration are common problems in the civic administrative systems and should be counteracted by using supplementary sources for the statistical Population Register.

7.1.1 Time references

The demographic events births, marriages and deaths are recorded together with the dates when they occurred. These time references are necessary if we want to produce population statistics for a specific point in time, which is reminiscent of the census week in a traditional census. The problem in many countries is migration. *From where, to where and when?* We must know when a residential address is valid to produce regional population statistics.

Assume that ten administrative registers are used to improve the quality of residential addresses in the statistical Population Register. If nine sources have an address A for a specific person and one source has another address B, which address should we choose? Trusting the majority is not a good method in this case. Instead, we should base the decision on time references – which address was valid at the reference time for the new register that we are creating? The conclusions from this example are:

- We should search for sources with time references. When was the register created and when was the residential address of each person recorded?
- The authorities responsible for administrative registers should be aware of the importance of time references.
- The authorities responsible for administrative registers should not overwrite old information. For example, old residential addresses are important when the NSO wants to study migration patterns.

7.1.2 Activities or 'signs of life'

The Birth Register for a specific year contains data regarding births, newborns, mothers and fathers as well as information about when and where the births took place. We can conclude from this information that the babies, the mothers and most of the fathers were active residents during the year.

In contrast, an administrative Population Register may include both active and inactive persons. Some persons can have been registered during many years but may have left the country long ago without reporting to the authorities.

Immigrants and mortality

Demographers outside Statistics Sweden complained that mortality rates for some immigrant groups were unreasonably low. After analysing several registers, we found that according to the longitudinal Income Register and the Population Register, 5.3% of the persons who were immigrants from the United States according to the Population Register had had zero income for more than eight years according to the Income Register. The corresponding value for Greek immigrants was 4.7% and for persons born in Sweden, the proportion was close to 0%. The total overcoverage in the Swedish Population register is between 0.5 and 1.0%, but as overcoverage is a very selective kind of error, this seemingly small overcoverage can destroy some estimates completely.

The conclusion from this example is that we must distinguish between registers with *persons* and registers with *persons' activities*. Chieppa et al. (2018) use this distinction when they estimate the population of Italy. There are *61 million* persons in their Population Register. This register can include overcoverage and can also be subject to undercoverage. They use several registers with persons' activities to find persons that are possible residents. Persons in the Population Register and at least in one of the registers with job and study activities, social benefits as retired, maternity allowance, unemployment benefits and tax return activities sum to *58 million* persons. This means that there are *61 − 58 = 3 million* persons in the Population Register that can be overcoverage as they are not included in any of the registers with activities.

There are also signs of undercoverage in the Population Register. About *1 million* persons have been active in the registers with job and study activities, but they were not included in the Population Register.

Conclusions – important sources

- Georeferencing the population requires sources with residential addresses and time references stating when persons moved to a new address.
- Reducing overcoverage and undercoverage requires sources with 'signs of life' or register activities.

Register activities can be used to delimit the resident population as follows:

1. *Persons who are in the Population Register and have activities/signs of life in other registers.* These persons probably belong to the resident population. The main rule should be to include this category in the register.
2. *Persons who are in the Population Register but have no activities/no signs of life in other registers.* These persons probably do not belong to the resident population; they are probably overcoverage. There is a choice between three rules: include all; exclude all; or include some according to a criterion. This

problem is discussed in Chapter 8 in connection with estimation methods using weights in registers.

3 *Persons who are not in the Population Register but have activities in other reg-isters.* These persons probably belong to the resident population; they are probably undercoverage. Subject matter considerations should define which categories to include. The rule should define permanent residents and some kinds of temporary residents studying or working in the country as members of the population.

7.2   The Population Register based on full information

In a traditional census, the interviewers collect data from all persons and households in the country. Identity numbers are not used for statistical processing and time references are easy – the conditions during the census week are measured during the interviews. With the short form in the census, detailed population statistics regarding persons and households for small regions can be produced.

The first step towards a register-based census is to use the Population Register for statistical purposes. As noted above, the Population Register is not only a base register. It is also an important survey, as the short form in the census can be replaced by information in the Population Register. This register is based on administrative data that is continuously being updated by the national registration system.

Perhaps the differences in survey methodology are most pronounced when we compare methods used for population statistics based on the census short form with methods used to produce population statistics based on a Population Register. When the statistical registers required for these population estimates are created, time references and identity numbers of persons and dwellings become essential.

7.2.1   Object types – Changing and unchanging registers

The Population Register contains information regarding four kinds of object types:

- *Persons*.
- *Demographic events* for persons.
- *Families* defined as married persons and their children, or persons with common children and their children.
- *Dwelling households* defined as the persons registered as permanently living in the same dwelling.

First, it is important to understand that 'the Population Register' at an NSO that produces register-based population statistics is not *one* register. It is a system of related registers that are used for several surveys. The system's base consists of the register with *all demographic events* that change the population of persons. The object set of persons can be changed and important character-istics such as place of residence can be changed by these demographic events.

The administrative Population Register is a living register that is undergoing continuous change. Newborns and deaths and other changes are registered when reported to the register authority.

Chart 7.2   The system of registers that constitutes the Population Register at the NSO

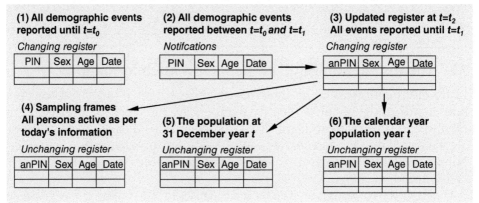

In Chart 7.2 the NSO starts with a copy of the administrative register (1) with all demographic events reported until t_0. The NSO then receives notifications (2) with new demographic events that have been reported to the administrative register for the period from t_0 to t_1. At the NSO, the NSO's register with all demographic events is updated at t_2 with this information (3).

The notifications are handled by the central unit responsible for receiving administrative data. At this central unit, the official PINs in the delivered data are replaced by the anonymous identity numbers (anPIN) used to protect con-fidentiality. With the updated register (3) containing the current stock of per-sons and households, *sampling frames* (4) are created for sample surveys that require data collection now.

When the NSO decides that almost all information regarding the previous year has been delivered to the NSO, the *standardised population* (5) with the population on 31 December is created. A *calendar year population* (6) can also be created with all persons that belonged to the population during the year or a part of the year. The registers (4)–(6) are statistical registers and are not changed after they have been created. All statistical registers are unchangeable registers unlike many living administrative registers that are constantly updated and changed with information about stocks of units (dwellings, enterprises, vehicles, etc.).

In Chart 7.2 three different versions of the statistical Population Register are illustrated by registers (4)–(6). More versions are discussed in Section 7.2.4.

7.2.2   Variables with different functions in the system
We differentiate between six types of variables in base registers such as the Population Register and Business Register, each with a different role in the system:

Chart 7.3   Different types of variables in the Population Register

anPIN	Address code	anPIN Mother	anPIN Father	anPIN Spouse	From_date	To register	To_date	Deregi-stration	Event	Sex	Region
2141	214	8469	1245	7252	1989 0301	1989 0420	null	null	3	2	1880
3244	310	1786	2376	null	2005 1101	2006 0108	null	null	3	2	2182
6229	134	4496	9926	null	1999 1201	2000 0115	null	null	3	1	1780
7048	339	3958	8441	6957	2010 1001	2010 1105	null	null	0	1	1981

anPIN	Name	Address
2141	Ppp Qqq	Aaa 1 Bbb
3244	Rrr Sss	Ccc 2 Ddd
6229	Ttt Uuu	Eee 3 Fff
7048	Vvv Www	Ggg 4 Hhh

Chart 7.3 continued

Addresses and names of persons should be protected. This information can be kept in a separate database with restricted access.

1. *Identifying variables* such as identity number anPIN, etc. are used to precisely identify objects. The corresponding IT term is *primary key*. An identifying variable should, if possible, be completely stable; that is, it should have the same value during the whole lifetime of the object.

2. *Communication variables* such as name, address and telephone number are used when the statistical office needs to contact an object regarding a questionnaire or an interview.

3. *Reference variables (foreign keys)* (Address code, anPIN Mother, anPIN Father, anPIN Spouse) are used to describe *relationships* between different objects. When matching registers contain data on different objects, reference variables produce matches between related objects.

4. *Time references* (From_date, To_register, To_date, Deregistration) are variables that specify a point in time for an event that affects objects or updates in the register. These variables are used when different register versions are created, such as the population at a specific point in time, and to describe the flow of demographic events during a given period of time.

5. *Technical variables* or variables for internal *register administration*. These variables often show the source or have comments on individual items or measurements. For example, the source for an enterprise's industrial classification code could be the Patent and Registration Office or the National Tax Agency. They can also be used to show which values have been imputed, correction codes or error codes. Variables with weights are used for estimation.

6. The actual *statistical variables* are used when data in the register are analysed and described. Certain variables, *spanning variables*, are used to define the cells in statistical tables. For every cell in a table, descriptive measures are calculated for other statistical variables, *response variables*. A variable can be a spanning variable in one context and a response variable in another context.

Standardised variables in the Population Register

Statistical variables that are used in many statistical registers and sample surveys should be stored and updated in the base registers. The names, formats and metadata for these standardised variables used for social statistics should be handled by the team responsible for the statistical Population Register. Sex, date of birth, region of residence, civil status, country of birth and degree of urbanisation are examples of variables in Statistics Sweden's Population Register.

A standardised variable is so important that responsibility for it is clearly defined. Only the team responsible for a standardised variable has the right to define names, formats and metadata. No one else at the NSO can use the name for other variables.

A register-based production system requires that all register teams at the NSO work together in a manner that promotes consistency and avoids double work.

7.2.3   Updating the Population Register
We assume that the Population Register at a statistical office for a small region consists of the data in Chart 7.4a regarding four persons at year-end 2012.

Chart 7.4a   The Population Register at December 31, 2012

anPIN	Address code	From_date	To_register	To_date	De-registration	Event	1 = Birth
2141	214	1989 0301	1989 0420	null	null	3	2 = Migrates from
3244	310	2005 1101	2006 0108	null	null	3	3 = Migrates to
6229	134	1999 1201	2000 0115	null	null	3	4 = Immigrates
7048	339	2010 1001	2010 1105	null	null	3	5 = Emigrates
							6 = Death

There are four variables with time references in the register:

- *From_date*: The time when the event happened; e.g. the person with anPIN 7048 moved to dwelling 339 on 1 October 2010 (date format *yyyy mmdd*. We recommend this date format as date then becomes an ordinal scale variable that can easily be used in register processing).
- *To_register*: The time when the Population Register was updated with the information regarding the event. On 5 November 2010, the register was updated with the information regarding the person with anPIN 7048.
- *To_date*: The time when the next event occurred regarding this person. The old situation ends at this date. Still active is coded as 'null'.
- *Deregistration*: The time when the observation is changed from active to inactive status in the register. Still active in the register is coded as 'null'.

Assume that the statistical office receives the notifications in Chart 7.4b regarding demographic events that have occurred since the previous delivery of information.

Chart 7.4b   Notifications regarding demographic events delivered 1 February 2013

anPIN	Address code	Date of event	Event	1 = Birth
2141	214	2012 1128	5	2 = Migrates from
3523	353	2012 1121	4	3 = Migrates to
6229	263	2012 1201	3	4 = Immigrates
7048	339	2012 1105	6	5 = Emigrates
9566	310	2012 1220	1	6 = Death

How should the Population Register be updated with this new information? The first step is to match the old version of the Population Register in Chart 7.4a with the new notifications. The linked observations are shown in Chart 7.4c.

Chart 7.4c Old register matched with the new notifications

anPIN	Address code	From_date	To_register	To_date	De-registration	Event	Address code	Date of event	Event
2141	214	1989 0301	1989 0420	null	null	3	214	2012 1128	5
3244	310	2005 1101	2006 0108	null	null	3			
6229	134	1999 1201	2000 0115	null	null	3	263	2012 1201	3
7048	339	2010 1001	2010 1105	null	null	3	339	2012 1105	6
Events:	1 = Birth		2 = Migrates from		3 = Migrates to		353	2012 1121	4
	4 = Immigrates		5 = Emigrates		6 = Death		310	2012 1220	1

The final updated register in Chart 7.4d contains all demographic events. New demographic events occur for three persons in the old register: one person emigrates, another leaves the old dwelling, and a third person dies.

One observation is kept as is (anPIN 3244). The three other old observations in the register are updated as shown in Chart 7.4d. Three new observations are added to account for person 6229 having moved into a new dwelling and two new persons having arrived, one through immigration and one baby was born.

7.2.4 Registers and time
Individual objects and object sets change over time. Objects are born, change location, are altered, or cease to exist. Different versions of the Population Register can be created with time references for these events. The processing of time references here is an important survey method. When defining a register with regard to *time*, the following register versions must be distinguished. The versions differ regarding the point in time or the period that the register describes.

1. The current stock register
The *current stock register* is the register version that is updated with all available information on currently active/live objects. The current stock register is used as a frame population for *sample surveys* or *censuses*.

Chart 7.4d Updated register 1 February 2013   **Chart 7.4e Three versions**

anPIN	Address code	From_date	To_register	To_date	De-registration	Event	Current stock 2012 1231	Population 2012 1231	Calendar pop. 2012
2141	214	1989 0301	1989 0420	2012 1128	2013 0201	5	1	0	1
3244	310	2005 1101	2006 0108	null	null	3	1	1	1
3523	353	2012 1121	2013 0201	null	null	4	0	1	1
6229	134	1999 1201	2000 0115	2012 1130	2013 0201	2	1	0	11/12
6229	263	2012 1201	2013 0201	null	null	3	0	1	1/12
7048	339	2010 1001	2010 1105	2012 1105	2013 0201	6	1	0	1
9566	310	2012 1220	2013 0201	null	null	1	0	1	1

In Chart 7.4d the current stock on 31 December 2012 consists of four observations as described by the first indicator variable in Chart 7.4e (grey-shaded).

For the current stock version at 2012 1231, the time *To_register* must be before 2013 0101 and the time for *Deregistration* must be after 2012 1231. Current stock versions are created for use as *frame populations* for sample surveys. They should not be used for register-based statistics, where register populations with better coverage should be used instead.

2. The population on 31 December

The *register referring to a specific point in time*, such as the end of the year, is the version of the register that is updated to describe the object set at that point in time. This update is carried out *after* the point in time, when information on all events up to that point in time is available. The update is used for *register surveys*.

Creating a register describing the actual population referring to a specific point in time requires more information than for the current stock version. If we assume that all information on events up to 31 December 2012 has been delivered to the statistical office on 1 February 2013, then the register version with the population on 31 December 2012 can be created. This population consists of four observations in Chart 7.4d, as described by the second indicator variable in Chart 7.4e.

For the register referring to 2012 1231, the time *From_date* must be before 2013 0101 and the time *To_date* must be after 2012 1231. The population referring to 31 December is used for most of the yearly register-based statistics on persons.

If a small number of births or a small number of deaths that occurred during 2012 are reported after the creation of the register referring to 31 December, this information should not be used to update the register. This is because the register is a standardised register used as register population for other surveys. Therefore, the register cannot be changed afterwards.

3. The calendar year register

The *calendar year register* in Chart 7.5 is the register version containing all objects that existed at any point during a specific year. Objects that are added or cease to exist during the year are included with information on the date of the event. It is used as register population for *register surveys*.

The calendar year register for 2012 consists of all persons that belonged to the population during 2012 or some part of 2012. Chart 7.4d shows that the complete register with both active and inactive observations actually is a register with all *demographic events*. But the current stock version 2012 1231 and the version referring to 2012 1231 are both registers of *persons*. However, the calendar version for 2012 contains seven observations for six persons and is therefore more difficult to handle. Later in the book we describe estimation methods for calendar year registers. Calendar year registers are particularly important for economic statistics.

The time *From_date* must be before 2013 0101 and the variable *To_date* must be after 2011 1231 for the calendar year register for 2012. Calendar year versions are important for flow variables such as income and production.

Objects that are added or cease to exist during the year are included with information on the date of the event.

Chart 7.5   Calendar year register for 2012

Object identity	Existed 1/1	Added	Ceased to exist	Existed 31/12	Other variables
PIN1	Yes	-	2012 0517	No	...
PIN2	Yes	-	-	Yes	...
PIN3	No	2012 0315	2012 0925	No	...
PIN4	No	2012 0606	-	Yes	...

4. The events register

In Chart 7.6 the *events register* for a specific period is the register containing information on all demographic occurrences that took place during the period. A register is created for every type of event. It is used in *register surveys*.

Chart 7.6   Events register for 2012 regarding change of address

Object identity	Address 1/1	Date of change of address	New address
PIN11	Address 11	2012 0517	Address 21
PIN12	Address 12	2012 0606	Address 22
PIN13	Address 13	2012 0911	Address 23

5. The historical register

The *historical register* in Chart 7.7 contains information on all demographic events that occurred at any time for each object. An object that has three events is shown on three rows, etc. It is used for *longitudinal surveys*.

Chart 7.7   Historical register regarding change of address

Object identity	From address	Date of change of address	To address
PIN21	Born	1967 0517	Address 1
PIN21	Address 1	1981 0606	Address 2
PIN21	Address 2	2012 0911	Address 3

6. The longitudinal register

A *longitudinal register* for a period of time (three years in Chart 7.8) is a register containing information on demographic events that took place during the period and the values for the statistical variables for all sub-periods (in this case, years) within the longer time period.

Chart 7.8   Longitudinal register for 2010–2012

PIN	Existed 1/1/2010	Added	Ceased to exist	Income 2010	Income 2011	Income 2012
PIN31	Yes	-	2011 0517	183 450	97 600	-
PIN32	Yes	-	-	273 500	281 360	258 340
PIN33	No	2011 0315	2012 0925	-	193 570	204 520
PIN34	No	2012 0911	-	-	-	56 300

7.2.5   Variables and time

Six different register types are discussed above, where time is treated in different ways. There are two types of variables where time is also treated differently. The combination of *variable type* and *register type* is important and should not be overlooked.

- *Stock variables* give the situation at a specific point of time, for example, age of an individual at a specific point of time or number of employees in an enterprise at year-end.
- *Flow variables* show sums for different time periods, for example, earned income during a year for a person and new orders during a month for an enterprise.

A stock variable should be defined for a register population defined for a specific point in time. A flow variable should be defined for a calendar year population. A calendar register with earned income of persons during a certain year should consist of all persons belonging to the population during at least some part of that year.

One version of the Swedish Income & Taxation Register (I&T) describes the income during year t of persons permanently living in Sweden on 31 December, year t. However, there are persons who were permanently living in Sweden earlier than 31 December, but who had left Sweden or died before 31 December. Their incomes would have been included in a calendar year register for year t. In the I&T Register describing 31 December, the total income is smaller than the total income in the calendar year register for year t. The National Accounts needs information regarding calendar year registers.

Example: The Swedish Income Register 2010
In the population on 31 December, 7.2 million persons earned SEK 1 821 billion during 2010. In the calendar year population for 2010, 7.4 million persons earned SEK 1 834 billion during 2010. The difference is due to persons leaving the population during 2010.

7.3   The Population Register in new register countries

Countries that want to create a statistical Population Register for the first time face a number of problems. The first tasks must be quality assessment of potential sources and the first version of the Population Register. Here we discuss the following issues:

- There can be different systems of identity numbers used by different authorities. What is the reason and what strategy should be chosen?
- In some countries there is no central Population Register for administrative purposes. But as a rule, there are many other registers with data on persons. How can these registers be used to create a statistical Population Register?

- The addresses in the administrative Population Register are not updated and are therefore of low quality. But there are other registers that can have information about addresses. How should the address variable be handled in the short run and long run?
- Section 5.5 discusses output data quality of new statistical registers and the quality of estimates produced with new registers. This discussion regarding the statistical Population Register continues in Chapter 8.
- Many sample surveys today have nonresponse rates close to 50%. But published estimates are based on methods that aim at reducing nonresponse errors. Can corresponding methods be developed to adjust for errors in registers? Estimation methods that can be used to reduce coverage errors in registers are discussed in Chapter 8.

7.3.1   Different systems of identity numbers

When work starts with creating a new statistical Population Register, the first thing that needs to be addressed is the issue of identity numbers. Many countries have established a unified system of national identity numbers, and these countries have good preconditions. However, there are countries with different kinds of numbers used in different parts of central and local administrations.

There can be one system with social security numbers linked with the national system of identity cards. However, hospitals and schools often develop their own systems since social security numbers are often given only to adults that have jobs. This makes record linkage difficult when registers with different identity numbers are combined. The last part of Chapter 4 has an example of a research institute that has serious record linkage problems due to hospital registers having developed their own system of identity numbers.

Health system identity number generates duplicates

In a particular country, the social security number and the identity card related with this identity number are well established and used by most citizens in their daily life. Hospitals can also use the social security number, but since some categories of patients do not have a social security number and the related identity card, the hospitals issue a simple paper card with a health system identity number for each patient. A patient who has forgotten to bring the paper card or who has lost it will get a new card with a new identity number. This will generate duplicates in the hospital registers. These duplicates cause problems for the administration of patients – the hospitals will not find earlier information for some patients. If the NSO uses registers from hospitals in their work with population statistics, then the duplicates may be difficult to find and eliminate.

The solution is to improve the health administrative system so that the social security numbers are used in the first place. The statisticians at the NSO should take the initiative to improve the identity number system in the country. The social security number should be assigned to all persons permanently living in the country including newborns – the social security numbers then become the national identification numbers.

If the statisticians at the NSO instead spend time to develop advanced methods to find duplicates and develop advanced record linkage methods, their efforts will not contribute to an efficient statistical system. Probabilistic record linkage will never cure the disease. If mismatch rates decline from, for example, 11% to 7%, this can be acceptable for making one record linkage operation. However, it is not sufficient in a production system for official statistics that utilises administrative registers.

7.3.2   Problems in countries without a central Population Register
Chart 7.9 gives a picture of the situation in a country without a central Population Register. The long-term goal should be to improve cooperation and coordination so that all registers in the second column are based on the central Population Register and that there is only one kind of identity number in the third column.

Chart 7.9   Many registers, but no register that covers the entire population

1. Authority/Ministry	2. Register	3. Identity number
Ministry of Health	Birth Register	No identity numbers
Ministry of Health	Death Register	National identity number
Social Security Board	Social security register	Social security number
Border Police	External migration	Passport number
Tax Department	Tax Register, persons	Tax identity number
Ministry of Health	Hospital registers	Health system identity number
Ministry of Education	School and university registers	Student identity number
Electoral Board	Electoral register	Voters card
Social Security Board	National id-card	Social security number

This situation is also described in Chart 2.10 in the section on *Many registers, but no good registers – what to do?* The situation is common in new register countries and leads to some difficult and important problems regarding statistical quality.

– The first column shows that many authorities are responsible for different parts of the statistical system, which causes the national statistical system to not be coordinated. FAO (2010) discusses integrated or coordinated national statistical systems. We have selected the following quotations from FAO's publication:

One of the shortcomings of current statistical systems in both industrialized and developing countries is that data are collected by sector, using different sampling frames and surveys. The division of data by sector leaves no opportunity to measure the impact of an action in one sector on another...
... In some cases, different organizations produce statistics for the same items, with different results, which confuse the data users.

– The transition to a register-based statistical system opens the possibility to develop a coordinated survey system where populations and variables in different surveys harmoniously fit with each other. As registers can be combined and sample surveys can be combined with registers, inconsistencies can be found by comparing populations and variable values from different parts of the system.

- Which combination of registers has the best coverage, and where are the best residential addresses with time references? A lot of work is required with quality assessment and consistency editing to answer these questions.
- How should the national system be improved? What authority should become the future national registration authority responsible for the central Population Register with the future national identity number? This is an important area for survey system design, which must be accomplished in the long term.

An inventory should be carried out to find two kinds of sources: sources that improve coverage and sources that improve residential addresses. Time references are also necessary to evaluate residential addresses.

Countries without a Population Register – how to create one?
None of the registers in Chart 7.9 are intended to cover the entire population of the country. The distributions by age and sex in these registers differ from the distribution for the whole population. But these registers contain 'signs of life'. They suffer from undercoverage as they do not cover the whole population, but there should be less risk of overcoverage. Creating a statistical Population Register of sufficient quality should be possible by combining several sources of this kind. For example, if the Electoral Register is combined with the Birth Register for the past 18 years, the combined register should include almost all adults and all children in the country.

A combined register of this kind can be the starting point for work with creating a statistical Population Register – in other countries with different preconditions, another combination can be chosen as the starting point. When the new register has been created, the next step will be to measure its quality. After that, perhaps more sources must be added; then quality is measured again.

7.3.3   How to improve coverage of the Population Register
In countries that have an administrative Population Register, the work with the statistical Population Register can focus on the following steps:

1. Analyse how the administrative register is handled
How is the work with the present civic register organised? There is a central office and regional offices – how do these cooperate? What routines and manuals control work with the register? Make an inventory of how data are structured – are there registers for different categories and different purposes?

Make an inventory of variables. What variables are important for the register authority? Compare with the variables in Chart 7.3 with time references. Are changes or updates appropriately documented? Analyse metadata for the register with the quality indicators A1–A9 in Section 5.4.1.

Does the register authority have sufficient and adequately trained staff for work with the register – at the central level and in the regions? Cooperation with the NSO is a new responsibility and improvements may be required to meet statistical needs. Developing and maintaining the Register of Residence noted in Section 2.7.1 will require strengthening of the authority responsible for the civic register.

2. Get delivery of the present register with the statistically important variables

After the first delivery of the civic register to the NSO, the statisticians will start work with analysing the microdata. The input data quality of the civic register should be analysed with the quality indicators B1–B7 in Section 5.4.2. The statistically important variables should be analysed. The experiences of the statisticians' work with quality assurance should be discussed together with the staff at the authority responsible for the civic register. Courses or study circles could be organised.

The delay in reporting births and deaths, marriages and divorces to the civic register should be analysed. The duration is measured between the point in time when the demographic event happened (From_date in Chart 7.3) and the point in time when the event was recorded in the register (To_register in Chart 7.3). This will determine how long a wait is required before the register population at a specific point in time can be created. When migration is reported to the civic register, the wait regarding migration should also be measured.

When a Population Register has been established at the NSO, how should it be updated with new information? An IT system must be developed that will make a new, updated version based on notifications of all changes in the population between the previous delivery and the current one. Recall that the basis for work with the Population Register at the NSO is the population of all *demographic events*. In addition, cooperation between the NSO and the civic authority relates to these demographic events. Creating the populations of *persons* used for official statistics is easy with this population of *events*.

In new register countries, the new Population Register at the NSO probably starts with yearly copies of the civic register and then gradually proceeds from quarterly, to monthly or daily updates. The updates should be based on the comparatively small amounts of data with only the new demographic events,[1] instead of deliveries of the whole administrative register.

3. Analyse and reduce undercoverage

For one particular country where the NSO has created the first version of a statistical Population Register, coverage was improved in the following way. After integrating more than 20 sources individually with the civic register and after elimination of duplicates, the NSO estimated the undercoverage of the civic register to be about 4.7%.

---

[1] Statistics Sweden receives between 10 000–15 000 notifications each day with demographic events relating to a population of about 10 million persons.

	Percent of final population
Civic register	95.31
Social security	1.88
Birth register	1.16
Vaccination	0.78
Social support	0.45
Education	0.11
Victim register	0.10
Health insurance	0.07
Health care	0.05
Handicapped	0.04
Other registers	0.05
TOTAL	100.00

When sources are added in this way, it is important that you do not add overcoverage. It is an advantage if the added sources are registers with active members of the population – the sources should contain 'signs of life'.

The added registers show errors in the civic register. These errors can be reduced and finally eliminated through more cooperation between the authorities responsible for all of these registers and after improving the routines used by the civic register.

After these improvements, the civic register could now be used alone as the central Population Register.

Section 2.5.2 discusses the case when there are many registers, but no good registers. The solution consists of cooperation and coordination. The NSO in this example has spent time on quality assessment of the above sources. This is an important learning process. The final register is also checked with macro-editing – the distribution by age and sex is compared with population projections.

4. Analyse and improve residential addresses
When the NSO in the example above integrated the more than 20 sources, they could also compare residential addresses in these sources. Choosing the most recent address requires knowing when each source was updated with the address information.

5. Analyse and reduce overcoverage
Emigration is an important cause of overcoverage – people leave without informing authorities. In some countries, data from the Border Police are used by the NSOs. Scanned passports give data on the identity of persons leaving and entering the country and when the border was passed. If a person has been outside the country for at least 6 or 12 months, the person can be classified as an emigrant. Copete, Sanchez and Recaño (2016) and Cruz and Cerrón (2015) have developed methods that use data from border passages to estimate international migration.

One conclusion in Section 7.1.2 is that reducing overcoverage requires sources with 'signs of life' or register activities. People who have a job, study, receive health care or receive social support, etc. are persons who we can assume are not overcoverage – they are active members of the population. Registers with such information should be used to create an indicator variable in the statistical Population Register describing 'register-activity'. This variable can be used to describe and improve quality by improving estimators. This is discussed in Chapter 8.

7.3.4   Inventory of sources – addresses and time references

This section continues the discussion in Sections 2.6 and 2.7. The traditional census aims at answering the central issues that are fundamental for social statistics. *Where do people live, and with whom?* Here we also note a third issue that is important in the census: *What is the housing standard of the households?*

In a register-based statistical system, the answers to these questions depend on the Population Register. The Population Register should be based on administrative systems that continuously generate data so that all persons' residences can be georeferenced, and the members of each dwelling household are georeferenced at the same dwelling.

Date stamping of these residences will place the population in both space and time. If reporting to a central register authority is mandatory when persons move to a new dwelling, the Population Register will contain the required information. We call this a Population Register based on full information in Section 7.2.

If persons can be linked to dwellings or residential addresses, they can also be linked to the real estate and buildings where they live. The yearly tax register used for property taxation contains variables that can be used to describe the housing standard of households.

Dias et al. (2016) investigated if the Real Estate Register in Portugal could be used for a register-based population and housing census. Countries that conduct a fully register-based census use the Real Estate Register in this way. However, in countries where the Real Estate Register is not sufficiently good, the housing part of the census can be replaced with a large-scale sample survey.

Countries without a Population Register with full information

The long-term goal is to achieve a system with full information regarding domestic migration. But what can be done in the short term? Many of the registers in Chart 7.9 contain information on residential addresses. There is also information about when the address information was recorded – the date of a hospital visit, the date of an application for social support, etc. A Population Register with a better quality of addresses can be created by combining a first version of a Population Register with residential addresses linked with time references from several sources. The next step is to measure the quality of these residential addresses with a sample survey.

Chart 7.10 outlines the first steps towards a new statistical system with a statistical Population Register. The *first step* could be to use the sources numbered 1 to obtain the coverage of the population and the sources numbered 2 to obtain better residential addresses. Of course, the sources numbered 2 can also be used to improve coverage, and some of the sources numbered 1 have information on residential addresses. The residential address is the link between the person and geography. Population statistics for small areas as municipalities can be produced with this link. But the addresses must have sufficient quality – we must know where people live now, not where they lived 10 or 20 years ago.

Chart 7.10   The first steps towards a new statistical system

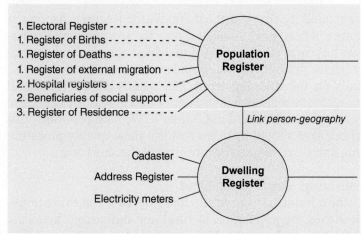

1. Electoral Register - - - - - - - - - - -
1. Register of Births - - - - - - - - - - -
1. Register of Deaths - - - - - - - - - - -
1. Register of external migration - -
2. Hospital registers - - - - - - - - - - -
2. Beneficiaries of social support -
3. Register of Residence - - - - - - -

Population Register

Link person-geography

Cadaster
Address Register
Electricity meters

Dwelling Register

Dwelling households cannot be created with residential addresses only. Persons belonging to many households may live at one address. To create households, there must be some administrative system that generates dwelling identities that can be linked to persons.

The *second step* is to develop this link between persons and dwellings. The Nordic countries have such systems. The Nordic statistical systems can link persons to the dwelling where they are registered as permanently living. And when people move to a new dwelling, reporting is mandatory to the authority responsible for the administrative Population Register.

But this improvement took 21 years for Norway and Sweden to accomplish. Quite a lot of work was required – addresses had to be standardised, dwellings had to be numbered, a dwelling register had to be created, legislation had to be changed and all persons in the population had to report at what dwelling they were living. Each dwelling is linked to the property, and the housing conditions for each household can be described with the Real Estate Register and the Property Tax Register.

Several countries do not have Real Estate, Property Tax registers or Cadastres that cover the entire country. Then it becomes necessary to give up the idea of describing housing conditions with these registers. These reduced ambitions make it possible to take *a shortcut for the second step*. The NSO in Ecuador has developed a method that uses electricity meters as a proxy for dwellings. Electricity companies have registers where these electricity meters are georeferenced. Almost all households are connected to a meter and the companies send bills via mail. The names on an invoice can be wrong; it can be a previous resident, etc. Nonetheless, the meter registers can still be used as a frame of the households in the country. During the census, interviewers can collect data regarding the meter identity number for each household together with identity numbers of the household members.

The *third step* towards a new statistical system should be to develop a Register of Residence (numbered 3 in Chart 7.10). Reporting to this register should be mandatory when persons move to a new dwelling, and the identity number of the old and new electricity meters should be included in these reports.

7.4 Methods to measure and improve quality

Assume that we have created the first statistical Population Register. As long as the time references have not been established, it is only for an approximate reference time. The register has three main statistical variables: age, sex, and municipality of residence. Can we publish regional population statistics based on this register? That depends on the coverage errors and quality of the residential municipalities according to the register. This was discussed previously in Section 5.5 and the discussion continues here and in Chapter 8. We have previously discussed long-term actions that should be taken to solve these quality problems. Here, we discuss actions that can reduce the problems in the short term.

7.4.1 Three kinds of surveys should be combined

What can be done if the administrative system does not cover the entire population? In several countries the population in rural and indigenous areas are not well covered by the civic registration. Many people also work in an informal sector – this informal economy is not illegal, but it is not included in the administrative systems in many countries.

In many cases, an administrative system covers only a specific part of the intended statistical population, and data must be collected from the part that is not covered using a sample survey. We distinguish here between situations when some categories of the population are not included at all in the system and the situation when important variables are not included in the administrative system.

Chart 7.11 illustrates the three different kinds of surveys discussed here: the register survey A, the register-based sample survey B, and the sample survey C based on an area frame. The area sample surveys have theoretically no coverage errors; therefore, they can be used to measure and reduce frame errors in other surveys, that is, surveys A and B.

Chart 7.11   Different parts of the target population and the desired variables

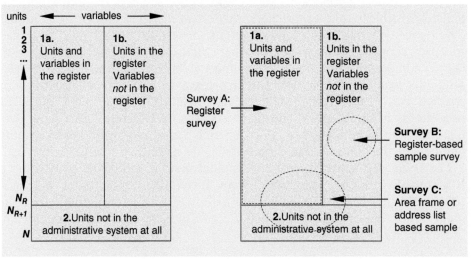

In Chart 7.11 the target population is divided into two parts: the objects or units that are included in the administrative register (1) and the objects that are not registered at all (2).

Here, two surveys should be combined: a register survey A describing part 1 and a sample survey C describing part 2 based on maps or address lists where data are collected by interviewers.

If the administrative register used for part 1 does not contain all necessary variables for statistical purposes, another sample survey B, using a frame created with the register (often called 'list' frame), can be conducted to collect data regarding these variables (1b).

The area or address list based sample survey for part 2 of the population can be used to give an estimate of the register's undercoverage. This sample will consist of two categories: objects that are included in the register and objects that are not included. It is necessary that the sampled units or objects can be identified with the same identifying variables that are used in the register. Combining a register and area sampling is an example of a *dual frame design*.

If the sample is a simple random sample, the size of the target population can be estimated with a simple ratio estimator. If we have a register with $N_R = 8\ 000$ objects and conduct a sample survey of $n = 1\ 000$ units from an area frame without undercoverage with a map or address list based sample, then the estimate is derived as illustrated in Chart 7.12.

Chart 7.12   Estimation of undercoverage

	In sample survey		Not in sample survey	Total target population		
In the register	y	e.g. 781	$N_R - y$ e.g. 7 219	N_R	e.g.	8 000
Undercoverage	$n - y$ e.g.	219		$\hat{N}_{UC}$	e.g.	2 243
Total	n e.g.	1 000		$\hat{N}$	e.g.	10 243

The ratio of register units in the sample is $p_R = y/n$ and can be used to estimate the corresponding population ratio N_R / N. This gives $\hat{N} = N_R / p_R$ and the size of the undercoverage is estimated as $\hat{N}_{UC} = \hat{N} - N_R$.

Ferraz (2015) considers a dual frame design where a list-based or register-based sample and an area frame based sample containing the same variables are combined during the estimation stage. All variables are collected for both samples. Dual frame designs are usually developed for situations when two random samples based on different frames covering the same population are combined.

We discuss a different case here that is important for countries that have started using administrative registers to create statistical registers. Coverage errors and problems with residential addresses that have not been updated may make it necessary to combine a statistical register with an area sample. The produced statistics are mainly based on the register or on sample surveys based on frames created with the register. With the register, producing estimates by municipality is desirable. The area sample is used to estimate coverage errors at national and regional levels and for different categories of municipalities and persons or households. Adjusting or correcting for coverage errors is also possible with calibration conditions based on variables that are present in both the sample survey and the register.

7.4.2   A new register-based system for statistics on persons

The main quality problem with the statistical Population Register in new register countries is caused by migration within the country. Chapter 8 discusses estimation methods that can be used when residential addresses are not updated when people move to a new address. New register countries without a Population Register with full information must have a strategy to handle this problem.

The new system for statistics on persons will be based on the statistical Population Register. The georeferencing of persons is based on the link PIN – geocode in the Population Register. Other register surveys and register-based sample surveys use this link to georeference persons and households. As a consequence, if this link in the Population Register is of low quality, then all surveys regarding persons and households will have the same quality flaws.

The quality of the new statistical Population Register must be measured. Chart 5.10 is repeated below, where we assume that we have conducted an area sample and the interviewers have collected data on where people live. This information is then compared with the residential information in the Population Register. The chart shows that there are errors in the regional classification. According to the register, 2 888 000 persons are classified as living in region A. However, according to the area sample, 16 000 + 2 000 of these are not living in region A. Also, 80 000 + 110 000 + 19 000 persons live in region A, but they are not classified as living in region A according to the register information.

Instead of using the term *classification error* for these errors, we use the terms *overcoverage* and *undercoverage*. The reason is that the 290 Swedish municipalities are important users of Statistics Sweden's regional statistics, which is why we can treat these regional statistics as 290 different register surveys. Chart 7.13 shows coverage errors in the register's three subpopulations A, B and C.

Chart 7.13   Coverage errors and area sample estimates of the population by district

From Chart 5.10		District according to register data			Not in register	All	Under-coverage
	District	A	B	C			
District	A	2 870 000	80 000	110 000	*19 000*	3 079 000	*209 000*
according to	B	16 000	3 110 000	40 000	*27 000*	3 193 000	*83 000*
area sample	C	2 000	10 000	1 530 000	*26 000*	1 568 000	*38 000*
	All	2 888 000	3 200 000	1 680 000	**72 000**	7 840 000	
	Overcoverage	*18 000*	*90 000*	*150 000*			

Persons not in the register at all can be found using an area sample. As long as these kinds of coverage errors in the Population Register cannot be neglected, it would be wise to continue with area samples to measure these coverage errors.

With a register-based sample survey (Chart 7.14), we can only measure regional misclassifications; we cannot measure the undercoverage of persons who are missing in the register. Persons who have left the country is another category of overcoverage that cannot be detected with the methods discussed here.

Chart 7.14   Coverage errors and register-based sample estimates of population

	District	District according to register data				Under-coverage
		A	B	C	All	
District according	A	2 870 000	80 000	110 000	3 060 000	*190 000*
to register-based	B	16 000	3 110 000	40 000	3 166 000	*56 000*
sample	C	2 000	10 000	1 530 000	1 542 000	*12 000*
	All	2 888 000	3 200 000	1 680 000	7 768 000	
Overcoverage		*18 000*	*90 000*	*150 000*		

7.5   Conclusions

In Section 5.1, four quality concepts are illustrated as shown in Chart 7.15. This chart is used here to illustrate the main quality issue facing new register countries: the georeferencing of persons in the statistical Population Register. If residential addresses are not updated, then the production of regional population statistics of sufficient quality will be difficult.

Chart 7.15   A register survey: From administrative registers to statistical estimates

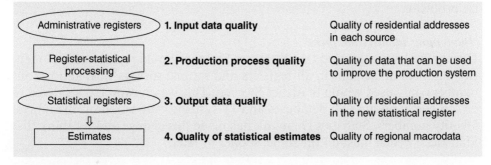

Administrative registers	1. Input data quality	Quality of residential addresses in each source
Register-statistical processing	2. Production process quality	Quality of data that can be used to improve the production system
Statistical registers	3. Output data quality	Quality of residential addresses in the new statistical register
Estimates	4. Quality of statistical estimates	Quality of regional macrodata

The Population Register is an important part of the administrative system and national statistical system, which is why great efforts should be spent on quality assessment and improvements.

Input data quality
The system of identity numbers can be improved, the civic register can be improved, and a Register of Residence can be developed so that people will start to report when they move from one dwelling to another.

Production process quality
Supplementary sources can be used to improve coverage and improve home addresses. Sources with time references can improve georeferencing, and 'signs of life' will reduce coverage errors.

Output data quality
The quality of the microdata in the statistical Population Register can be analysed with area samples or register-based samples as in Charts 5.10 and 7.13.

Quality of statistical estimates
Calibration of sampling weights is an important method to reduce coverage errors, sampling errors and nonresponse errors in sample surveys. In a similar manner, calibration of weights in registers can be an important method to reduce coverage errors in register surveys. This is discussed in Chapter 8.

Calibration of sampling weights has reduced nonresponse errors in the Swedish Labour Force Survey (LFS). The nonresponse rate is almost 50%, but after calibration of sampling weights the nonresponse error can be as low as 1%. Due to nonresponse, some categories are underrepresented, and some categories are overrepresented in the sample. Many auxiliary variables from registers are used, and if there are variables that are strongly correlated with nonresponse patterns, estimates using calibrated weights can theoretically be unbiased.

7.6   Challenges in old register countries

In this chapter, we have discussed the problems that new register countries will face. Old register countries that have much better preconditions, also face similar problems.

Anonymous identity numbers
During 2001 we suggested that Statistics Sweden should replace official identity numbers of persons in all registers and sample surveys with anonymous numbers only used within Statistics Sweden. This was, however, considered to be too costly because all databases and IT systems were based on the official identity numbers. Statistics Denmark has made this change and we think that Statistics Sweden must do the same, sooner or later.

Undercoverage
Statistics Sweden's surveys with data on persons have defined their frames and register populations so that only persons registered as permanently living in Sweden are included. This makes statistics production easy, but as an increasing number of persons come to Sweden for studies and work, the problems with undercoverage are becoming increasingly serious. Many of these persons from abroad are registered as temporarily living in Sweden, and their study

and job activities are recorded in university and tax registers that could be used by Statistics Sweden.

Since the Population Register at Statistics Sweden is used to create frame populations for the Labour Force Survey and other surveys, these sample surveys inherit the same coverage errors.

Wrong residential addresses

Young persons who leave their parents' home and travel to study at a university often forget to report to the administrative Population Register that they have moved. About 1% of the population have the wrong residential address in the register; but among young people, about 16% have the wrong residence. The solution should be to use more administrative sources such as the University Register to improve the statistical Population Register.

CHAPTER 8

The Population Register – Estimation Methods

Summing up Chapters 1–7: A register survey is conducted by using administrative registers and the system of statistical registers to create a new statistical register. The next step is to use the new register's data matrix or matrices to produce the relevant estimates and statistical tables for specific research objectives.

We discuss some quality problems associated with the Population Register and give suggestions for solutions to these problems based on certain estimation methods. These estimation methods are based on the principle that weights are used for register-based statistics in a similar way as for sample surveys.

When the data matrix is used to create statistical tables, the table cells will contain frequencies, sums or other statistical measures. When weights are used for estimation, weighted frequencies or weighted sums are calculated.

Because the registers in the register system interact, coverage errors, missing values and other quality problems in one register will affect other registers which import data from it. Therefore, the methods we propose must function within the whole system so that the statistics from different registers are consistent.

Calibration of weights – basics
This chapter describes first the method of calibrating and using weights in a register with a small example with register data (Section 8.1).

Calibration – a real application
Thereafter, the calibration of weights in the Swedish Labour Force Survey is explained. Section 8.2 shows how sample surveys in a register-based system benefit from auxiliary variables in the register system and provides an advanced example of how calibration can be done in a real situation.

Calibration to reduce coverage errors
Section 8.3 returns to the problem of bad quality residential addresses and shows how calibration of weights in registers can be a method that reduces coverage errors.

Register-based Statistics: Registers and the National Statistical System, Third Edition. Anders Wallgren and Britt Wallgren.
© 2022 John Wiley & Sons Ltd. Published 2022 by John Wiley & Sons Ltd.

Using weights to reduce overcoverage
Finally, Section 8.4 returns to the problem with overcoverage of persons in the register with no register activities or 'signs of life'.

8.1   Estimation in sample surveys and register surveys

The term *estimation* is generally used for sample surveys, but it should also be used with register-based statistics. Distinguishing between the actual values in the target population and the estimates produced by the register is also important here.

Statistical inference in sample surveys consists mainly of methods for point estimates. These point estimates should be as good as possible; unbiased estimators with small variances are preferable.

How can these concepts be transferred into the subject field of register surveys? With sample surveys, estimates for domains are made using formula (1) below. The design weights d_i depend on how the sample has been designed or allocated into different strata. The weights g_i in formula (1) are based on the auxiliary variables from statistical registers and are used to minimise sampling errors and errors caused by nonresponse.

$$\hat{Y} = \sum_{i=1}^{r} d_i g_i y_i = \sum_{i=1}^{r} w_i y_i \qquad \text{where } r \text{ is the number of units in the } sample \text{ that responded in a particular domain.} \qquad (1)$$

Deville and Särndal (1992) introduced this method of estimation where the original weights d_i are replaced by the *calibrated weights* w_i. The book by Särndal and Lundström (2005) explains how calibration of sampling weights can be used to reduce nonresponse bias in sample surveys.

No special methods are currently used when register-based statistics are produced; instead, calculations and summations are made in the simplest possible way:

$$\hat{Y} = \sum_{i=1}^{R} y_i \qquad \text{where } R \text{ is the number of units in the } register \text{ in a particular domain.} \qquad (2a)$$

We interpret these seemingly simple calculations as estimates. The values of these estimates depend on the methods used when the register was created. If this work is carried out in different ways, there will be different numerical values in the register-based statistics produced with the register. Choosing the methodology for the creation of a register means also choosing an estimation methodology.

In register surveys, all design weights are $d_i = 1$, as all observations in the register population or domain are included in the summation. Formula (2a) can then be written as in (2b) to clarify that all observations have equal weights:

$$\hat{Y} = \sum_{i=1}^{R} d_i y_i \qquad \text{Where all } d_i = 1 \text{ and } R \text{ is the number of units in the } register \text{ in a particular domain.} \qquad (2b)$$

8.1.1   Estimation methods for register surveys that use weights

In addition to the fundamental estimation methods that are determined by how the register is created, this chapter introduces weights w_i to solve some of the quality problems. The weights are calculated in different ways for different problems. These weights make possible correcting for different types of errors, e.g. that the register estimates are on an incorrect level due to coverage errors. Estimates are made here by using formula (3):

$$\hat{Y} = \sum_{i=1}^{R} d_i g_i y_i = \sum_{i=1}^{R} w_i y_i \qquad \text{Where all } d_i = 1 \text{ and } R \text{ is the number of units in the } \textit{register} \text{ in a particular domain.} \tag{3}$$

With traditional methods all $g_i = 1$, but other weights will be used in this chapter. The types of errors we discuss in this chapter include errors due to item nonresponse or missing values and overcoverage or undercoverage. The methodology could be used for more kinds of errors.

8.1.2   Calibration of weights in register surveys

This section illustrates how weights d_i can be calibrated by an example based on the register in Chart 8.1. Of the 19 observations in the register, two have missing values, observation 6 and 15. Four persons are not employed and therefore have no industry code, but these are not missing values.

Chart 8.1   Register of all persons in two small districts

(1) PIN	(2) Sex	(3) District	(4) Employed	(5) Industry	(6) Education	(7) d_i	x_{1i} Sex=F	x_{2i} Sex=M	x_{3i} District=1	x_{4i} Employed=1	w_i
1	F	1	0	null	Low	1	1	0	1	0	0.98276
2	M	1	1	A	Low	1	0	1	1	1	1.15517
3	F	1	1	A	Low	1	1	0	1	1	1.13793
4	M	1	1	A	Medium	1	0	1	1	1	1.15517
5	F	1	1	A	Medium	1	1	0	1	1	1.13793
6	M	1	1	*Missing*	Low	0	0	1	1	1	0.00000
7	F	1	1	D	Medium	1	1	0	1	1	1.13793
8	M	1	1	D	High	1	0	1	1	1	1.15517
9	F	1	1	D	Medium	1	1	0	1	1	1.13793
10	M	1	0	null	Medium	1	0	1	1	0	1.00000
11	F	2	0	null	Low	1	1	0	0	0	1.00000
12	M	2	1	D	Low	1	0	1	0	1	1.17241
13	F	2	1	D	Low	1	1	0	0	1	1.15517
14	M	2	1	D	Medium	1	0	1	0	1	1.17241
15	F	2	1	D	*Missing*	0	1	0	0	1	0.00000
16	M	2	1	A	Low	1	0	1	0	1	1.17241
17	F	2	1	A	Medium	1	1	0	0	1	1.15517
18	F	2	1	A	Medium	1	1	0	0	1	1.15517
19	M	2	0	null	Medium	1	0	1	0	0	1.01724

If we want to estimate a frequency table describing education by industry with this register, the missing values will affect the estimates. The table in Chart 8.2 is based on the yellow columns in Chart 8.1 and simple summations with the weights d_i.

Chart 8.2a   Employed persons by education and industry

Number of persons	Industry A	Industry D	Industry A %	Industry D %
High education	0	1	0.0%	16.7%
Medium education	4	3	57.1%	50.0%
Low education	3	2	42.9%	33.3%
All	7	6	100.0%	100.0%

Chart 8.2a differs from standard practice in many countries, where missing values are often customarily shown as 'Industry unknown' and 'Education unknown' as in Chart 8.2b. Missing values are shown in this way, but the estimates have not been adjusted to reduce errors due to the missing values. Grey-shaded cells are the same in both charts.

Chart 8.2b   Employed persons by education and industry, missing values are shown

Number of persons	Industry A	Industry D	Industry unknown	Industry A %	Industry D %
High education	0	1	0	0.0%	14.3%
Medium education	4	3	0	57.1%	42.9%
Low education	3	2	1	42.9%	28.6%
Education unknown	0	1	0	0.0%	14.3%
All	7	7	1	100.0%	100.0%

Charts 8.2a and 8.2b show that missing values give rise to estimation problems and problems regarding how data are presented to users. Here, this example is used only to discuss calibration of weights.[1]

The variables in columns (2), (3) and (4) in Chart 8.1 have no missing values. These variables can be used to calibrate the weights d_i so that estimates using the calibrated weights w_i will be adjusted for the missing values in columns (5) and (6).

Sums and/or frequencies based on the variables without missing values can be used as *calibration conditions*. There are many ways to choose these; and each choice will give calibrated weights that can differ. In this example, we use four conditions:

Calibration conditions	Redundant information (not used)
The correct number of women = 10, of men = 9, of persons in district 1 = 10 and of employed = 15	Total number of persons = 10 + 9 District 2 = 9 Not employed = 4

---

[1] In Chapter 12 in Wallgren and Wallgren (2014), there is more on missing values in register surveys.

This means that we use the marginal distributions for the variables sex, district and employment status as calibration conditions. If these four frequencies are estimated with the set of observations with missing values, the weights d_i should be used. The estimates will be erroneous due to missing values:

The number of women = 9 (error = –1)	The number of men = 8 (error = –1)
The number of persons in district 1 – 9 (error = –1)	The number of employed – 13 (error – –2)

The idea with calibration is to adjust the weights d_i so that the errors of these four estimates will be zero. All other estimates will also be adjusted in the same manner. Using the new weights, consistent estimates can be produced that have been adjusted for the missing values in the register.

The first seven columns in Chart 8.1 show the original register, while columns $x_{1i} - x_{4i}$ contain the auxiliary variables to be used when calibrating the weights.

In the calculations, x_i' vectors are used, one vector per row. For $i = 1$, such as for the person with identity PIN = 1, the vector $x_i' = (1\,0\,1\,0)$.

The summations are now referring to all observations in the register, not only one domain or cell as in the earlier formulas (1)–(3). The last column in Chart 8.1 shows the calibrated weights w_i, that are calculated in the following three steps:

1. $T = \sum d_i x_i x_i'$ and T^{-1} are calculated, where all $d_i = 1$ (missing values, $d_i = 0$) and $i = 1, 2, ..., 19$.

T is a matrix with sums of squares and product sums, here a 4×4 matrix:

$$T = \begin{bmatrix} \sum d_i x_{1i}^2 & \sum d_i x_{1i} x_{2i} & \sum d_i x_{1i} x_{3i} & \sum d_i x_{1i} x_{4i} \\ \sum d_i x_{2i} x_{1i} & \sum d_i x_{2i}^2 & \sum d_i x_{2i} x_{3i} & \sum d_i x_{2i} x_{4i} \\ \sum d_i x_{3i} x_{1i} & \sum d_i x_{3i} x_{2i} & \sum d_i x_{3i}^2 & \sum d_i x_{3i} x_{4i} \\ \sum d_i x_{4i} x_{1i} & \sum d_i x_{4i} x_{2i} & \sum d_i x_{4i} x_{3i} & \sum d_i x_{4i}^2 \end{bmatrix}$$

2. The vector λ is calculated: $\lambda = T^{-1}(t_x - \sum d_i x_i)$.

The vector t_x is the four conditions for the number of women and men, persons in district 1, and persons employed.
The vector $\sum d_i x_i$ is the corresponding unadjusted number.

t_x	$\sum d_i x_i$	$t_x - \sum d_i x_i$
10	9	1
9	8	1
10	9	1
15	13	2

The vector t_x represents the correct values of the four calibration conditions, and the vector $\sum d_i x_i$ represents the erroneous values based on the observations with missing values.

3. The adjusted weights become: $w_i = d_i(1 + x_i'\lambda)$. The adjusted weights are used to calculate weighted numbers and totals.

These formulas are illustrated below, where the calculations are done step by step.

1. The matrices T and T^{-1} are calculated:

$$T = \begin{bmatrix} 9 & 0 & 5 & 7 \\ 0 & 8 & 4 & 6 \\ 5 & 4 & 9 & 7 \\ 7 & 6 & 7 & 13 \end{bmatrix}$$

$$T^{-1} = \begin{bmatrix} 0.375000 & 0.250000 & -0.125000 & -0.250000 \\ 0.250000 & 0.362069 & -0.112069 & -0.241379 \\ -0.125000 & -0.112069 & 0.237069 & -0.008621 \\ -0.250000 & -0.241379 & -0.008621 & 0.327586 \end{bmatrix}$$

2. The vector λ is calculated:

$$\lambda = \begin{bmatrix} 0.375000 & 0.250000 & -0.125000 & -0.250000 \\ 0.250000 & 0.362069 & -0.112069 & -0.241379 \\ -0.125000 & -0.112069 & 0.237069 & -0.008621 \\ -0.250000 & -0.241379 & -0.008621 & 0.327586 \end{bmatrix} \cdot \begin{bmatrix} 1 \\ 1 \\ 1 \\ 2 \end{bmatrix}$$

$$\lambda = \begin{bmatrix} 0.000000 \\ 0.017241 \\ -0.017241 \\ 0.155172 \end{bmatrix}$$

3. The adjusted weights become: $w_i = d_i(1 + x_i'\lambda)$

For the first person in the register, $i = 1$, and $x_1' = (1 \quad 0 \quad 1 \quad 0)$

$$x_1'\lambda = \begin{bmatrix} 1 & 0 & 1 & 0 \end{bmatrix} \cdot \begin{bmatrix} 0.000000 \\ 0.017241 \\ -0.017241 \\ 0.155172 \end{bmatrix} = -0.017241$$

The calibrated weight for person 1 becomes: $w_1 = 1 \cdot (1 - 0.017241) = 0.98276$

The calibrated weights for the other persons are calculated in the same way and are in the last column in Chart 8.1.

The weighted frequencies in Chart 8.3 are estimated with calibrated weights. The relative frequencies differ from the frequencies in Chart 8.2a, and the number of employed persons now sums up to 15 (15 = 8.1 + 6.9) instead of 13.

Chart 8.3 is a better alternative than the complicated presentation in Chart 8.2b. In Chart 8.2b the problem on how to interpret the effects of missing values is handed over to the user – in Chart 8.3 this problem has been solved by the statistician.

Chart 8.3   Persons by Education and Industry, adjusted for missing values

Weighted number of persons	Industry A	Industry D	Industry A %	Industry D %
High education	0.0	1.2	0.0%	17.4%
Medium education	4.6	3.4	56.8%	49.3%
Low education	3.5	2.3	43.2%	33.3%
All	8.1	6.9	100.0%	100.0%

8.2   Calibration of weights – the Swedish LFS

The sampling and estimation procedure in the Swedish Labour Force Survey is described in Statistics Sweden (2013). The nonresponse bias in the LFS is analysed in Statistics Sweden (2017).

Chart 8.4 illustrates the rapid growth of nonresponse rates by age group. Nonresponse rates differ nowadays significantly between young and old, and also between sexes, regions, levels of education and country of birth.

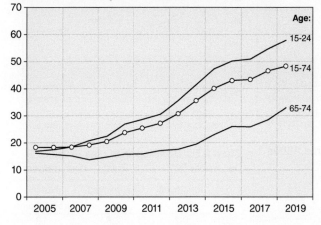

Chart 8.4   Nonresponse rates in the Swedish LFS

Due to these selective response patterns, some categories are underrepresented in the response set and other categories are overrepresented.

This is similar to the situation faced by new register countries. Due to underreporting of migration within the country, declining regions are overrepresented, and growing regions are underrepresented in the Population Register

8.2.1   Use of auxiliary information in the LFS

The sampling and estimation procedures in the Swedish Labour Force Survey (LFS) can use hundreds of potential register variables as auxiliary information for stratification and estimation. Powerful auxiliary variables that are strongly correlated with *response propensity*, important *study variables* and *domains* can give almost unbiased estimators even if the nonresponse rate is high. The book by Särndal and Lundström (2005) provides a solid background for calibration of weights.

The LFS uses register variables for stratification and calibration of weights:

Strata are defined by: Sex • Age-group • Region = 2 • 3 • 24 = 144 strata

And 66 auxiliary variables are used for calibration of weights:

Sex • Age-group = 2 • 13 = 26 categories from the Population Register
Region   = 26 categories from the Population Register
Country of birth   = 4 categories from the Population Register
Employment by industry = 8 categories from the Employment Register
Unemployment   = 2 categories from the Register of Unemployment Support

8.2.2   Nonresponse bias in the LFS

Information in administrative registers can be used to determine Labour Force Participation for those who belong to the LFS frame population. This information has good quality but is available only two years after the survey. If this information is used for the nonresponse set in the LFS, the parameters in the LFS can be estimated without disturbing nonresponse errors. The nonresponse bias can be estimated after comparisons with the published estimates. In Chart 8.5 the nonresponse bias is illustrated in charts describing estimated trends for the main LFS variables.

Chart 8.5 Nonresponse bias in the LFS, December 2011–2015. Relative bias, percent

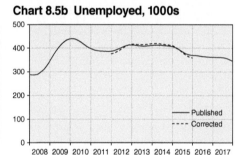

Chart 8.5a Employed, 1000s

Chart 8.5b Unemployed, 1000s

Chart 8.5c Not in Labour Force, 1000s

Chart 8.5d Nonresponse bias in the LFS
December 2011–2015. Relative bias, percent

	Employed	Unemployed	Not in Labour Force
2011	0.7	2.8	−1.8
2012	1.0	−0.8	−2.0
2013	1.0	−1.9	−1.9
2014	1.3	0.9	−2.6
2015	1.1	2.9	−2.7

The nonresponse bias is analysed in Statistics Sweden (2017). The main estimates regarding Labour Force Participation for persons 16–74 years are shown in Chart 8.5d. Charts 8.5a–8.5c describe how the bias disturbs the time series patterns for the main variables. The disturbances are small.

    The main conclusion in Statistics Sweden (2017) is that there is nonresponse bias in the LFS. The level of the relative bias differs between 0–15% for aggregate estimates. Since there are auxiliary variables correlated with labour force participation and variables as sex, age and country of birth, the relative bias is small for domains defined by these variables.

The variable *level of education* was not used as an auxiliary variable. For domains defined by this variable, the relative bias regarding employment was between 10–20% (negative bias for low level of education and positive bias for high level of education).

Note that the nonresponse bias did not increase when the nonresponse rate increased during the period.

The nonresponse rate in the Swedish LFS is currently about 50%. This means that the *input data quality* is low for the microdata collected by interviewers. However, if these microdata are combined with register variables from Statistics Sweden's system of statistical registers, the *quality of the statistical estimates* is usually high. The register variables are used for stratification and calibration of sampling weights. Statistics Sweden's LFS publishes every month about 15 000 point estimates based on these calibrated weights.

8.3   Calibration – where do people live?

Residential addresses are often not updated in administrative population registers in many countries in Latin America. This means that using administrative population data is difficult for regional statistics. In countries with strong urbanisation, this will lead to overcoverage in rural register populations and undercoverage in urban register populations. A statistical Population Register is important for rural and agricultural statistics. It is, for example, important to know if poor farmers have given up farming and moved to a city. For all these reasons, using a statistical population register for population statistics at the national level is not sufficient – regional population statistics are perhaps even more important.

This section presents ideas on how estimates based on statistical registers can be adjusted for coverage errors with calibrated weights. The adjustments are made in a similar way as when sample survey estimates are adjusted for nonresponse errors. Here, however, we adjust the register-based estimates with a sample survey instead of adjusting the sample-based estimates with a register. This method for adjusting register estimates for coverage errors can be used during the transition period when administrative systems are gradually being improved so that coverage errors are reduced.

- First, coverage errors in the statistical register should be reduced by using all relevant sources.
- Secondly, the coverage errors should be monitored with area sample surveys, as in Chart 7.13.
- A third step could be to adjust the register-based estimates with calibrated weights that are based on a comparison between the area sample and the register. A register-based sample can also be used, as in Chart 7.14.

Regional register-based statistics in many Latin American countries are difficult because migration within the country is not recorded in the national registration systems. To test our ideas regarding calibration of weights, we have used linked microdata from Statistics Sweden's Population register for 2005 and 2015 regarding persons who belonged to the population in both these years. This means that the only demographic events we study are migration within Sweden.

Coverage errors in the Population Register can be caused by different problems with the registration system, and each problem should be tackled with a specific long-term project. Underreporting of births and deaths should be reduced by improving registration methods. Underreporting of immigration and emigration should be reduced through cooperation with migration authorities and improved border control.

However, the most serious problem is migration within the country. The final solution is to change the national registration system so that reporting to the register becomes mandatory when a person changes residential address. But this requires new legislation and changed attitudes, which will take time. In the short term, the methods we present here can be used where estimates are adjusted for net migration.

Assume that the Swedish register for 2005 represents a bad register from a Latin American country that has not been updated regarding migration within the country. We call data from 2005 'Registro Civil' not updated for migration, and we call data from 2015 the 'unknown Truth today'. We use data from 'Registro Civil' combined with sample survey data from the register for 2015 to estimate the 'unknown Truth today'. Migration creates undercoverage and overcoverage in regional statistics.

Microdata from 'Registro Civil' can be compared with the 'Truth' for the Swedish municipality Åsele. The population of Åsele is 619 + 2 081 = 2 700 persons according to 'Registro Civil', but only 2 081 + 305 = 2 386 persons according to the 'Truth'. The gross error due to coverage errors is 619 + 305 = 924, but the net error is only 619 – 305 = 314. Only net errors (net migration) are important for the estimates.

Overcoverage, 23%	619	} 2 700, Population according to 'Registro Civil'
Correct domicile	2 081	
Undercoverage, 13%	305	} 2 386, Population according to the 'Truth'

Adjusting for coverage errors due to net migration requires auxiliary variables in the register that are correlated with net migration. The 290 municipalities in Sweden are grouped into ten categories ranging from the category with the highest degree of urbanisation that should have the strongest net migration *into* the municipality to the category with lowest degree of urbanisation that should have the strongest net migration *from* the municipality. Countries in Latin America can find variables that can be used to create similar categories by analysing the last two censuses and using information in other available registers. Returning to the Swedish example, we also want to try age as an auxiliary variable as young persons have a higher propensity to migrate than old persons.

The first estimation exercise, *Exercise 1* in Chart 8.7, does not try to improve estimates from 'Registro Civil'. In *Exercise 2*, 'Registro Civil' has been supplemented with microdata from another administrative source that is common in Latin America. Poor people who seek support go to a regional office and fill in a form that includes information on the person's identity number and present place of residence. If a person moves to another municipality, the person must update the information in the poverty register (called Sisben in Chart 8.7) to receive continued support. Since about half of the population in Latin American countries can be included in this kind of poverty register, the coverage errors in Exercise *2* are reduced by about 50%.

The first calibration exercise (*Exercise 3* in Chart 8.7) uses only one auxiliary variable, *category* of each person's municipality of residence. We postulate that no special sample survey is necessary for calibrating weights in the statistical Population Register, since already existing household surveys based on area sampling can be used. The only necessary information to include is each person's identity number and the municipality of residence in the survey. Exercise 3 requires only ten estimates from the sample survey and the corresponding estimates from the register. The ratios between these estimates are the calibrated weights in Chart 8.6. This simple case does not require the matrix algebra in Section 8.1.

Chart 8.6   Population estimates, Exercise 3

Category	'Registro Civil'	Area sample	weight
1	351 107	317 396	0.90399
2	268 733	249 748	0.92935
3	340 143	320 935	0.94353
4	385 406	369 257	0.95810
5	577 247	558 744	0.96795
6	765 929	748 678	0.97748
7	936 860	929 009	0.99162
8	1 011 517	1 014 497	1.00295
9	2 037 960	2 080 777	1.02101
10	1 288 365	1 374 226	1.06664
All	**7 963 267**	**7 963 267**	

Exercise 3 uses weights in the 'Registro Civil'. The 290 municipalities have been grouped into ten categories, from Category 1 with expected strongest net migration out from the municipalities up to Category 10 with the expected strongest net migration into the municipalities.

The weights are computed as the ratios between good population estimates from the area sample and bad estimates from 'Registro Civil'.

In *Exercise 4* the 'Registro Civil' is improved with information in the Sisben Register. The quality of the estimates is now quite good. If the 'Registro Civil' that has not been updated during ten years is supplemented with information on residence in the poverty register and if the estimation uses weights derived from the simple model in Chart 8.6, then the municipality population estimates are reliable. In the Sisben part of the register, persons have weights close to 1; but in the remaining part of the register the weights vary between 0.90 and 1.07.

Chart 8.7 compares the 290 estimates regarding the population of each municipality with the true values.

Chart 8.7   Exercises 1–5: Reducing coverage errors in population estimates

Exercise:	Mean absolute error, %	3rd quartile	Maximum absolute error, %
1. 'Registro Civil' (RC), not updated for migration,	4.4	6.4	16.8
2. RC supplemented with Poverty Register (Sisben)	2.2	3.2	9.0
3. RC, Calibrated weights, 10 categories	0.6	0.6	6.0
4. RC and Sisben, calibrated weights, 10 categories	0.4	0.5	3.3
5. RC and Sisben, cal. weights, 10 cat. and age groups	0.4	0.5	3.8

Exercise 5 also uses age group as an auxiliary variable together with the categories used in Exercise 4. With seven age groups combined with ten categories, the area sample is used to estimate 70 parameters.

Chart 8.7 clearly shows that the variable age does not improve the quality of the estimates of the municipality populations; however, the age distributions of the persons in the municipalities are clearly improved. According to Chart 8.8, the number of persons 20–39 years with high propensity for net migration are increased by the weights for Stockholm municipality, but are decreased for the rural Åsele municipality.

Chart 8.8   Estimates for Exercise 1 ('RC') and Exercise 5, percent of population, %
Stockholm   municipality   (Category   10)   Åsele municipality (Category 2)

Age	'RC'	Ex. 5	Truth	Error ex.1	Error ex.5	Age	'RC'	Ex. 5	Truth	Error ex.1	Error ex.5
10–19	11.6	10.6	10.3	1.4	0.3	10–19	10.7	11.2	11.1	-0.4	0.1
20–29	13.4	15.8	17.0	-3.6	-1.2	20–29	14.6	10.3	11.2	3.3	-0.9
30–39	15.5	17.6	17.9	-2.4	-0.3	30–39	10.4	7.8	8.6	1.8	-0.8
40–49	17.7	16.3	15.6	2.1	0.7	40–49	10.5	10.7	10.5	0.0	0.2
50–59	15.3	14.8	14.7	0.7	0.1	50–59	14.3	15.3	15.1	-0.8	0.2
60–69	12.8	12.3	12.1	0.8	0.2	60–69	16.1	18.2	17.8	-1.7	0.4
70 +	13.7	12.8	12.5	1.1	0.2	70 +	23.5	26.6	25.6	-2.1	0.9
All	100.0	100.0	100.0			All	100.0	100.0	100.0		
Pop.	670 474	717 621	718 008			*Pop.*	2 700	2 433	2 386		
Error %	**-6.6**	**-0.1**				**Error %**	**13.2**	**2.0**			

Combining an area sample and a register
Our conclusion from these estimation exercises is that in Latin American countries, regional statistics based on a statistical Population Register can be improved with the methods recommended here. We have used real Swedish data with errors that we think are similar to what countries in Latin America experience. Statistics for small regions were improved using a simple model with two auxiliary variables.

Combining an area sample and a register-based sample

In a traditional dual frame approach, one list-based sample survey is combined with one area sample. The surveys are conducted simultaneously, and the same variables are collected in both sample surveys. We suggest a completely different method for register-based sample surveys:

- The area sample is only used to adjust the Population Register.
- The register-based sample uses this adjusted Population Register as a sampling frame. The calibrated weights in the Population Register are included in the data matrix for the sample.
- The sample estimates are calculated with sampling weights that are the design weights adjusted for nonresponse, multiplied with the calibrated weights from the Population Register. The sample estimates are then adjusted for both nonresponse and coverage errors.

Sample surveys carried out this way will be consistent with the Population Register. Moreover, all other statistical registers with data on persons should use the adjusted Population Register as the register population. Since the same weights are used, these registers and sample surveys will be a coordinated system with consistent statistics.

8.4   Methods to handle overcoverage

The Population Register may include active residents and, for instance, persons who have left the country without reporting to authorities. Emigration should be measured primarily by, for example, border control officers when people leave the country. In Section 7.3.3 we note that some countries have developed methods that use data from scanned passports for this purpose. Such methods can reduce new additions to the overcoverage. However, emigration has been extensive in some countries for many years. Then it may be necessary to estimate the size of the overcoverage with methods based on absence of 'signs of life' in registers with activity data.

Macro-editing is important

Unreasonable estimates can be found in register-based statistics by being observant during macro-editing and analysis. The question should be asked whether overcoverage could be the cause of these extreme estimates.

  The first sign of overcoverage in Statistics Sweden's Population Register came from demographic studies on mortality. Mortality was strangely low among a few categories of foreign-born persons. In addition, the average income for those born in different countries could be misleading. For persons born in certain countries, the negative bias was found to be around 20%.

Use available sample surveys

If overcoverage is suspected, available sample surveys can be used to measure the overcoverage. When the National Statistics Office (NSO) sends questionnaires to persons in a sample, the postal staff returns mail addressed to persons they could not find. In a similar manner, the NSO's interviewers should report persons they did not find. As this information comes from sample surveys, estimating overcoverage is possible for different categories of respondents. Nonresponse in the Swedish LFS has been used to estimate overcoverage in some categories of foreign-born persons in the Swedish Population Register.

Use registers with 'signs of life'

By combining information in several registers, persons in the Population Register can be divided into three categories: clearly active residents, clearly inactive and the doubtful rest who are difficult to classify. The clearly inactive persons should be removed from the statistical Population Register by the responsible staff. The doubtful category can be used together with weights as illustrated below.

Weights can be used to adjust for the estimated overcoverage. Before adjustment, all weights are equal to 1. After adjustment, the weights for the different categories for which there is overcoverage will be less than 1. All other statistical products using the Population Register will then use these weights. Thus, all the statistics produced will be consistently corrected for the estimated effect of overcoverage.

The following method builds on ideas developed by the National Statistical Institute of Spain; see Argüeso and Vega (2013).

1. The clearly inactive are removed from the register. The register population then consists of two categories: clearly active and doubtful.

2. The doubtful category should be analysed to provide an understanding of the nature of the overcoverage: Age? Nationality? Province of residence?

3. The register population is divided into homogenous classes by, e.g. age, nationality and province of residence. Homogeneity here means similar probability of belonging to the overcoverage. Each class contains C_i clearly active persons and D_i doubtful persons. Of the doubtful, only a part, $w_i \cdot D_i$, should be included in the population estimate; the rest are assumed to be overcoverage. The weights w_i should be estimated with an area sample. In the register there are T_i observations in class i:
$$T_i = C_i + D_i$$
and the true number of individuals in class i is:
$$N_i = C_i + w_i \cdot D_i$$

4. The area sample is linked with the register. The persons who have been interviewed in the sample survey are all active residents according to the interview. We obtain estimates from the sample survey of the real number N_i of persons who belong to each class or subpopulation in the register. From the estimated N_i and the known C_i and D_i, we can obtain the weights w_i:

$$w_i = ((t_i - c_i)/c_i)/(D_i/C_i)$$ where t_i and c_i are estimates of T_i and C_i from the sample $$\hat{N}_i = C_i + w_i \cdot D_i$$

In the final register, all clearly active persons are assigned the weight 1 and all doubtful persons in class i are assigned the weight w_i. By applying these weights, estimates of population statistics can be produced that have been adjusted for overcoverage. We can also find persons in the area sample who are not included in the register. Then we can obtain estimates of the undercoverage in each class or subpopulation of the Population Register.

How to define the doubtful category

The doubtful category is defined as persons that do not have any activities or 'signs of life' in a specified selection of administrative sources that we will use as indicators of overcoverage. The problem is that persons can have no activities in these registers but still belong to the population.

Assume that we have a Population Register with 10 000 persons, of whom 5% have left the country without reporting to the authorities. Then there are 500 observations in the register that are overcoverage. The sample survey has given us this estimate. The doubtful category can be defined in different ways. Here we compare different definitions:

1. No sign of life in any of the indicator registers during the last two years.
2. No sign of life in any of the indicator registers during the last three years, etc.

The first attempt could use the indicator registers in this way and try to find information about the 500 persons in the register that can be suspected of being overcoverage. The most likely can then be removed. If this attempt is difficult, then the overcoverage can be handled in the way suggested by Argüeso and Vega.

CHAPTER 9

Defining Register Populations – Coverage Errors

The focus in previous chapters has been on social statistics or statistics on persons and the Population Register. The table in Chart 9.1 compares social statistics with economic statistics – the table defines important differences between the Population Register and the Business Register. When creating these two registers, the register populations should be defined differently due to these differences. The question is: how should such base registers as the Population Register and the Business Register be designed and used? These two registers at Statistics Sweden will be used as cases to illustrate topics of general interest.

Register populations can be defined in different ways and different sources can be used. These methodological issues affect overcoverage and undercoverage in the created registers. Because the Population Register and the Business Register are used to define populations in many surveys, coverage errors in these base registers will be transferred to other surveys.

Chart 9.1 Methodological differences between social and economic statistics

	Social statistics	Economic statistics
1. Main variables	Stock variables that exist for points in time, for example age, sex	Flow variables that exist only for periods, for example turnover, value added
2. Definition of target population	All units that existed at a specific point in time	All units that existed during any part of a specific period (year, quarter, month)
3. Permanence of population units	Persons are the same units during their lifetime, only their variable values change	Units can merge with other units or split into new units during the period
4. Equality of population units	All persons are of equal interest	A few big units are of main interest; many small units are sometimes disregarded (cut-off sampling)
5. Combining sample surveys and registers	Frame populations and register populations are very similar	Frame populations and register populations are created at different points in time and differ considerably in coverage

Register-based Statistics: Registers and the National Statistical System, Third Edition. Anders Wallgren and Britt Wallgren.
© 2022 John Wiley & Sons Ltd. Published 2022 by John Wiley & Sons Ltd.

Some similarities also exist between the Business and Population Registers. In several countries, a costly *Economic Census* could be replaced by administrative registers just as the Population Register makes it possible to replace the Population Census. A Business Register based on administrative sources, a number of administrative registers with economic information from different tax forms, and some sample surveys can be used to generate economic information that can replace the Economic Census.

The *Farm Census* is another important survey that can be replaced by a *Farm Register* based on administrative sources. Consistency will be easier to obtain if the Farm Register is included in the Business Register.

9.1   Defining a register's object set

For every statistical survey, the survey population needs to be defined. This section discusses the definition of the object set or population. We reserve here the concept *population* to refer to an object set that belongs to a specific survey. We use the concept *object set* when we describe a register without referring to a specific survey.

When creating a new register for a specific survey, the new register's population must be defined. Every source register has its own object set that will be included either completely or partially in the new register. However, every statistical register is created for one or several principal uses or surveys. It is therefore common that the register's object set agrees with the main survey's population.

9.1.1   Defining a population

Every survey begins with a set of questions formulated in theoretical or general terms. The theoretical concepts in the set of questions must be operationalised, i.e. translated into measurable concepts. Defining these measurable concepts determines *what* is being surveyed. A population should be defined as follows:

> **Defining a population**
> The population definition should clearly show which objects are included in that population. The *object type* should be specified. In addition, a *time reference* and *geographic delimitation* should always be included. The geographic delimitation should also specify the relation that exists between the objects or statistical units and the geographical area.

Example of a Definition for the Population Register:
'Permanently resident individuals in Sweden on 31 December 2021. Permanently resident refers to ...'

This definition includes the following components:

individuals	= object type
permanently resident	= relation
in Sweden	= geographical area
on 31 December 2021	= point in time

Example of a Definition for the Business Register
'Legal units classified as active in Sweden according to the Tax Board on 28 November 2021. To be classified as an active legal unit the unit should be registered for VAT and/or registered as an employer for Pay As You Earn (PAYE) and/or registered for F-tax (preliminary taxation of sole traders).'

This definition includes the following components:

Legal units	= object type
registered for tax	= relation
in Sweden	= geographical area
on 28 November 2021	= point in time

Sample survey theory and the guidelines for quality concepts and the quality declarations by statistical offices usually contain three concepts related to populations:

- *Population of interest* refers to the population in the theoretical question at hand.
- *Target population* refers to the operationalised population: the theoretical population of interest which has been translated into a concrete and examinable population, i.e. the population that is the *target* for the survey.
- *Frame population* refers to the object set that the frame actually generates. Data collection is planned from all or some objects in the frame population. A frame can consist of other objects than those to be surveyed, i.e. a map or an address register for a survey on households.

These three concepts can be found in the theory of sample surveys and censuses with their own data collection. For register surveys, we only use the two former concepts. The third concept, *frame population*, must be replaced because the sampling frame does not exist with register surveys:

- *Register population* refers to the object set in the register that has been created for the survey in question, i.e. the population that is actually being surveyed.

Important differences exist between a frame population and a register population (Chart 9.2). A frame population is defined *before* the data collection, while a register population is created *after* the reference period when all administrative data have been received.

Chart 9.2   Frame populations are created before register populations

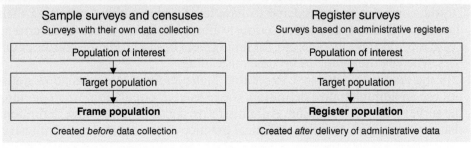

The structure of the population

Here, we want to survey the size and structure of the Swedish population on 31 December of a specific year. The *population of interest* is permanent residents in Sweden on 31 December. However, this vague concept needs to be further defined, and it can be defined in different ways. In general, a good definition should meet the following requirements:

- it should be *adequate*, i.e. in accordance with what you want to survey;
- it should be *functional*, i.e. it should be applicable in a practical sense.

When setting up a definition, finding the balance is often difficult between what you want to survey (adequate definition) and what is possible to survey (functional definition). In this example, the following definitions of *permanent residents in Sweden on 31 December of year t* could be possible:

- Definition (1): Persons registered by the National Tax Agency in Sweden on 31/12 of year t.
- Definition (2): Persons registered by the National Tax Agency in Sweden on 31/12 of year t, according to data available at the end of January of year t + 1.

Those working with the Swedish Population Register usually wait until the end of January to create the register version that relates to the situation on 31/12. Then the hope is that all the changes and events affecting the population register for the previous year have been reported. The created register should then be the applied version even if some notifications referring to year *t* can still be received. Therefore, the *register population* is defined using Definition (2) for this survey.

However, the quality declaration for the Swedish annual population statistics states that the survey aims to describe the *target population* according to Definition (1) given above. The difference between the target population and the register population is therefore the register's *coverage error*. Deaths and emigrations that have not been reported before the end of January cause *overcoverage*, while births and immigrations that have not been reported cause *undercoverage*.

In addition to these coverage errors, relevance error can occur if the definition is not adequate. The difference between the population of interest and the target population is one of the survey's *relevance errors*. Between 25 000 and 50 000 persons registered in the population register in Sweden probably do not live permanently in Sweden. About 4–8% of immigrants from outside the Nordic countries are considered to have left Sweden without reporting their departure. This relevance error seriously affects statistical estimates describing death rates, average income, etc. for immigrants from outside the Nordic countries, whereby the estimates become misleading.

In the example above, we can see that Statistics Sweden's population statistics use an administrative definition, the registered population, when defining the survey's target population. Administrative concepts always give definitions that are functional. Sensible statistical practice dictates that these administrative concepts should be used to define the target population if the relevance errors are small. However, the basic rule is that the population definition should attempt to meet the requirements of the statistical survey. If the administrative concepts are not sufficiently relevant or adequate, then developing own definitions and carrying out the required register processing will be necessary so that the register's object set reflects the defined population as closely as possible.

An improved definition

Is it possible to depart from the present administrative definition of Sweden's population? This is discussed within Statistics Sweden, and there are ways to improve the definition of the target population:

- Include foreign students studying at Swedish universities; the administrative data are available. These students are registered by the universities and have a special identity number assigned by the National Tax Agency.
- Include foreigners working in Sweden; the administrative data are available. The National Tax Agency has assigned them with a special kind of identity number. During 2009, 57 905 foreigners were found in the register of all Income Statements who were not found in the Population Register according to Laitila, Wallgren and Wallgren (2012). The fraction of undercoverage among the population of all employed persons in the Employment Register was 1.4%.
- Exclude Swedish students studying abroad; they are registered by the National Tax Agency, but do not live permanently in Sweden. Administrative data with information on persons who study abroad are available.
- In many cases, Swedish university students may be registered by the National Tax Agency where they resided before going to the university. Their present address is registered by the universities. If this information is used for the Population Register, then regional population statistics will be different and improved.

This means that we can depart from the administrative definition of the target population and introduce a new, more adequate statistical definition. When Statistics Sweden's Population Register was first established in 1967, these coverage problems were small – very few foreigners were studying or working in Sweden. Moreover, very few Swedish young people were abroad and still registered with their parents. Thus, coverage errors were small at that time. But today, these coverage errors are becoming increasingly disturbing. A change would be preferable, but this requires revision of a complicated IT system that handles the updating – and this is something that delays improvements.

Changed population definition changes coverage errors

With Definition (2) above, the coverage errors are small – only a small number of births, deaths and migration movements are missing in the register population. The differences between the target population in Definition (1) and the register population in Definition (2) are small.

If the definition of the register population is changed according to the improved definition above, then the relevance errors will be reduced, but the coverage errors may increase.

9.1.2   Can you alter data from the National Tax Agency?

The example above of the Population Register illustrates an attitude that has a long tradition in the Nordic countries. When register-based statistics were introduced during the 1960s, administrative data were used in many cases without changing object sets and object types. The Population and Business Registers were exact copies of the corresponding registers at the National Tax Agency. Even today there is some reluctance to making changes in administrative data. This attitude is also explained by the fact that methodologists as a rule have not taken part in the development of register-based statistics. There is a tradition among methodologists to create and change estimators and estimates, an attitude that is also needed for the production of register-based statistics.

The traditional attitude

As someone responsible for a register at a statistical office, should you change administrative data that come from another authority? The administrative authorities collect the data and therefore have primary responsibility for the register. A person working at a statistical office with a specific product or survey may not believe that the data should be changed or supplemented to suit the statistical purposes of the product in question.

The transformation principle in Chart 1.15 points out that administrative registers should be processed so that the objects and variables correspond to the statistical requirements. This means that the staff at a statistical office who receive the administrative data have both the freedom and the obligation to carry out such changes so that the quality of the statistics can be improved. The persons who make these changes should be experienced, independent and have the support of a network of register statisticians. Otherwise, they might not dare make the changes.

Not making any changes in the administrative data you receive can be very convenient. Then you can say that the administrative authority is responsible for all errors. If you make changes in the received data, you may feel that you are responsible for all errors. However, if you are a statistician, you are always responsible for the statistics you produce.

9.1.3   Defining a population – the Farm Register

The first version of the Farm Register was created by Statistics Sweden in 1968. The introduction of this register signified a great improvement in Swedish agricultural statistics. According to Jorner (2008), the quality was so improved that instead of being one of the worst in Europe, Swedish agricultural statistics became one of the best. The register was designed for both administrative and statistical purposes and was based on registers developed by farmers' associations and local agricultural boards.

The Farm Census and agricultural sample surveys used the Farm Register as frame, and questionnaires could be sent to all farms or a random sample of farms – data collection became easy and cheap. Today, the Farm Register is based on registers used for the EU's support and control of agriculture.

How should those working with registers define the target and register populations? Statistical registers are based wholly or partially on administrative registers; thus, there is a risk that the administrative system's object sets will influence the choice of register population inappropriately.

The object set in the administrative register may not completely cover the target population that is of statistical interest. The administrative object set consists solely of those objects that are included in the administrative system, and there can be both overcoverage and undercoverage compared with the statistically desirable target population.

The administrative object set – is it suitable as the target population?

This example illustrates a general problem. Sometimes an administrative system does not cover the entire country; rural parts of the country and areas with informal economy may not be included. In these cases, register surveys must be combined with sample surveys. Selander et al. (1998) propose such a combination for the agricultural example below.

Agricultural statistics are currently based on applications for subsidies that farmers in the European Union submit to the county administrative boards. These applications are registered in the IACS system (Integrated Administration and Control System), which is used to administer the agricultural subsidies.

Chart 9.3   Undercoverage in an administrative register

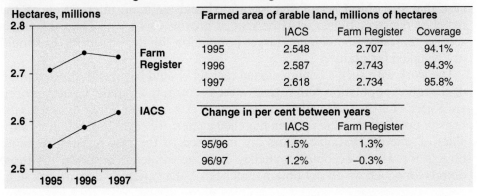

Farmed area of arable land, millions of hectares			
	IACS	Farm Register	Coverage
1995	2.548	2.707	94.1%
1996	2.587	2.743	94.3%
1997	2.618	2.734	95.8%

Change in per cent between years		
	IACS	Farm Register
95/96	1.5%	1.3%
96/97	1.2%	−0.3%

Chart 9.3 shows a comparison of area data in the subsidy applications within the IACS system and the corresponding data from the Farm Register, which was based on a census carried out by Statistics Sweden. The differences between IACS and the Farm Register are due to undercoverage in the IACS register – some farmers do not apply for subsidies even though they are actively farming.

Conclusions: Although the IACS register can be considered to have good coverage, seemingly small variations in the coverage mean that the time series for farmed area of arable land is misleading – an actual decrease in 1996–1997 appears to be a continued increase in the IACS register.

Flaws in coverage in administrative registers should not be ignored. On the contrary, target populations in these cases should be defined according to statistical requirements. Then a new register should be created containing the intended register population using the current base register – in this case by selecting units with economic activity ISIC = 01 from the Business Register. This new register can then be matched against the IACS register, enabling detection of any overcoverage or undercoverage in the IACS register.

Overcoverage is an indication of possible flaws in the Business Register. Undercoverage in the IACS register will appear as missing values in the new register. This 'nonresponse' can be corrected by adding a special survey to collect data from the part of the target population not included in the IACS register.

9.1.4   Defining a population – integrated registers

Integrated registers have been created by combining a number of registers. How should those working with integrated registers define target populations and register populations? We discuss the basic principles below.

Register commissions with matching – selection of target population
Many register commissions involve the combined processing of several registers. A series of matchings are carried out and variables are imported from different registers. The end result is an integrated register with many variables that are of interest to the project's customers.

But how has the survey's target population been defined? It can easily be the case that the object set of the integrated register is an intersection of the matched registers' object sets (the shaded area in Chart 9.4). Does this intersection represent an adequate target population for the project? This should not be taken for granted.

The work with such a commissioned register should begin with the definition of the target population with regard to the problem to be studied. An appropriate object set is then selected from the relevant base register. This object set is the register population, which is then matched against the registers containing variables of interest. Variable values are imported to the new register for those objects for which we obtain matches. Item nonresponse is shown for those objects for which we do not obtain matches, i.e. the variable values are missing.

Chart 9.4   Object sets when matching two registers

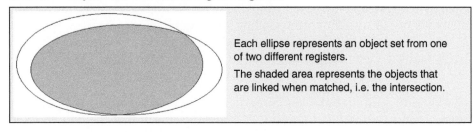

Each ellipse represents an object set from one of two different registers.

The shaded area represents the objects that are linked when matched, i.e. the intersection.

Conclusions: Which target population corresponds to the intersection? The intersection can never be a target population, as it would not be possible to give a definition of the intersection. Nor should it be the register population, as missing values due to non-matches would risk being forgotten. Being aware of the missing values is important, as is carrying out adjustments based on information regarding the scope and structure of the missing values.

9.2   Defining a calendar year population

Section 9.1 discusses a register referring to a specific point in time (31 December). A calendar year register is a different kind of register, where the population is defined as suitable for economic statistics, for example.

In Section 7.2.4, six different register types are discussed where time is treated differently. The combination of variable type and register type is important and should not be overlooked; this is explained in Section 7.2.5. There are two types of variables where time is also treated differently:

– *Flow variables* show sums for different time periods. Flow variables are relevant for calendar year registers. Both the population and the variables will refer to the same period of time.

– *Stock variables* give the situation at a specific point of time.

This section discusses simple examples where weights should be used to produce estimates for register-based statistics.

The calendar year register is the register version containing all objects that have existed at any point during a specific year. In a calendar year register, objects can exist during different time periods. Most objects do not cause problems, as they exist during the whole year. However, some objects are born or enter the register at certain moments and other objects die or disappear during the calendar year. This gives rise to estimation problems that can be solved by using weights. Time can be used as a *weight-generating variable*, and correct estimates can be produced for calendar year registers using these weights.

Average population
The average population in a municipality can be estimated as shown in Chart 9.5, where we calculate the day of birth or arrival in the municipality as a full day and the day when the person moved/died as no day in the municipality.

Chart 9.5   Calendar year register for the population in a (small) municipality

Person	Existed 1/1 2013	Arrived during 2013 yyyymmdd	Ceased during 2013 yyyymmdd	Existed 31/12 2013	Weight = Time in the municipality, years
PIN1	Yes	-	20130517	No	136/365 = 0.37
PIN2	Yes	-	-	Yes	365/365 = 1.00
PIN3	No	20130315	20130925	No	194/365 = 0.53
PIN4	No	20130606	-	Yes	209/365 = 0.57
Total	2			2	2.47

The traditional way of calculating the average population for 2013 is to form the average value of the population on 1 January 2013 (2 persons) and the population on 31 December 2013 (also 2 persons). A more specific calculation, in which time in the municipality is used as weight, gives the average population during 2013 as 2.47 persons instead of the traditional measure of 2.

Flow and stock variables
The data in Chart 9.6 relate to enterprises in a particular region during 2013. Certain enterprises began or ceased to exist at different times during the year. Time can also be used here as a weight-generating variable.

Flow and stock variables should be treated differently. Flow variables, such as the value added of an enterprise, only relate to the values during the period of the year in which the enterprise was active, and therefore do not need to be weighted. A stock variable showing the level at a point of time, such as number of employees, must be weighted. The total value added in the region during 2013 was SEK 83 million, while the average number of employees was 112.5. Productivity is calculated as $83/112.5 = $ SEK 0.738 million per employee per year.

Chart 9.6   Calendar year register for 2013 for enterprises in a particular (small) region

Enterprise identity	Existed 1/1	Started	Ceased	Existed 31/12	Weight	Value added	No. of employees	Weight • No. of employees
BIN1	Yes	–	20130630	No	0.50	10	30	0.50 • 30 = 15.0
BIN2	Yes	–	–	Yes	1.00	42	45	1.00 • 45 = 45.0
BIN3	No	20130401		Yes	0.75	31	70	0.75 • 70 = 52.5
Total					2.25	83		112.5

9.2.1   Defining a population – frame or register population?
During November each year, a frame with the current stock at mid-November is created by Statistics Sweden's Business Register. This frame is intended to be the basis for all yearly economic statistics. A number of economic surveys are based on samples from this frame. However, administrative registers are also used for economic statistics. The combination of frame-based sample surveys and register surveys based on administrative object sets highlights important methodological issues.

We formed a project group with members from the National Accounts, the Business Register and the main economic surveys. The charts below are from the final report from this project (Statistics Sweden, 2007b). Two methods were compared in the project:

– The traditional method where the *November frame* for 2004 was used for the Structural Business Statistics survey (SBS). The Swedish SBS is a combination of an administrative register with yearly income declarations from enterprises and a sample survey to some enterprises to collect more detailed information.

– A new approach where a *calendar year register* was created. This register was based on all administrative sources for 2004 available at the end of 2005.

Chart 9.7 illustrates the differences between the two approaches. The times when the quarterly sampling frames were created are marked with arrows above the time axis. The November frame was created approximately one year before the calendar year register. The inflow of administrative data is shown under the time axis. There are 12 deliveries of monthly tax reports for 2004 and three deliveries during 2005 of annual tax reports for 2004. When the calendar year register was created, much more information was available. Thus, there must be coverage errors in the November frame that are not present in the calendar register.

The *target populations* are the same: the population of all enterprises that were active during some part of 2004. However, the November frame was based on the current stock version of the Business Register with enterprises active during November. Enterprises active during January–October or December but not active during November were not included. The reason is the assumption that questionnaires cannot be sent to 'dead' enterprises. An additional assumption was that if the November frame is used for all yearly economic statistics, then the statistics produced would be consistent.

The *register population* of the calendar year register was defined as all legal units that were active (= reporting not only zeros) in at least one administrative source for 2004. Trade data from Customs, VAT data, monthly wage sum reports, yearly wage sums and yearly income declarations were used.

Chart 9.7   The November frame and the Calendar Year Register 2004

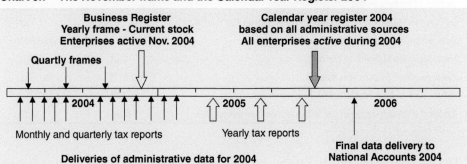

Chart 9.8 shows the overcoverage and undercoverage of the November frame measured as number of enterprises (legal units). Statistics Sweden's Business Register was at that time only based on one source – the Business Register at the National Tax Agency. The quality of the administrative variable activity in this source is clearly not sufficient for statistical purposes.

Chart 9.8   Over- and undercoverage as number of legal units in the November frame

November frame 2004			Calendar year register (CYR) 2004
93 114	'Active' according to November frame **Over coverage**	Not in CYR	
779 277	'Active' according to November frame	In CYR	779 277
47 662	'Has never been active' acc. to Nov. frame		
88 227	'Not active' according to Nov. frame	In CYR	307 577
171 688	Missing completely in Business Register		
307 577	**Total undercoverage**		

2005

15 Nov. 2004   15 Jan. 2006

The coverage problems are mostly generated by small enterprises. In Chart 9.9 the total effects of undercoverage on the estimates of wage sum and turnover are 0.64% and 2.13%, respectively. In Chart 9.10 the undercoverage is broken down by industry for the non-financial sector. We find that undercoverage is substantial for some industries.

Chart 9.9   Errors due to undercoverage in the November frame 2004, SEK million

	Wage sum	Turnover
'Active' according to the November frame	1 003 186	5 582 374
'Has never been active' according to the November frame	177	1 734
'Not active' according to the November frame	5 671	112 363
Missing in the November frame	639	5 061
Calendar Year Population, total	1 009 673	5 701 532
Errors due to undercoverage in the November frame	6 487	119 158
Errors due to undercoverage in the November frame, %	*0.64*	*2.13*

The overcoverage in the Structural Business Statistics survey (SBS) is treated as nonresponse, and positive values for economic variables are imputed. The estimates in Chart 9.11 are therefore subject to overcoverage errors that are small at the total level, but between 2.2% and 4.7% for some industries.

Some important conclusions can be drawn from this example:

– The National Tax Agency's Business Register should not be the only source for Statistics Sweden's Business Register. Instead, *all relevant sources should be used* (the transformation principle in Chart 1.15) to create a register population with good coverage.

Chart 9.10   Undercoverage in November frame 2004, non-financial sector by industry

Non-financial sector	Wage sums, yearly income statements			Turnover, VAT Register		
	Total	Undercoverage		Total	Undercoverage	
ISIC	SEK million		*Per cent*	SEK million		*Per cent*
01	4 663	341	*7.3*	64 720	4 425	*6.8*
10 and 11 and 12	178	14	*8.0*	1 399	24	*1.7*
212	2 988	2	*0.1*	26 094	4 441	*17.0*
23	1 040	1	*0.1*	11 761	1 455	*12.4*
702 except 70201	4 661	338	*7.2*	95 009	3 536	*3.7*
91	1 776	14	*0.8*	5 327	801	*16.0*
...	...	...	*...*	...	...	*...*
Total:	650 102	5 629	*0.9*	5 453 021	116 324	*2.1*

Chart 9.11   Overcoverage in SBS based on November frame 2004, by industry

	SBS, Number of legal units			SBS, Value of production		
	Total	Overcoverage		Total	Overcoverage	
Non-financial sector				SEK	SEK	
Industry, ISIC:			*Per cent*	millions	millions	*Per cent*
05	1 514	106	*7.0*	1 204	43	*3.5*
741	55 123	5 570	*10.1*	77 787	1 705	*2.2*
851	19 358	1 543	*8.0*	43 417	1 687	*3.9*
852	921	54	*5.9*	1 796	85	*4.7*
91	1 364	82	*6.0*	6 445	187	*2.9*
93	27 617	2 260	*8.2*	13 483	351	*2.6*
...	...	...	*...*	...	...	*...*
Total:	777 793	59 773	*7.7*	3 554 910	21 604	*0.6*

– The monthly administrative sources can be used to improve coverage of the quarterly frames.

– A calendar year register should be created the year after the calendar year in question. This register can be a standardised population for all sources used by the yearly National Accounts.

– Undercoverage can be a very selective error that cannot be neglected. Note that in Chart 9.10, the errors due to undercoverage affect wage sums and turnover differently – different industries have different errors. This means that these variables will be inconsistent even when they are included in the same survey, as in the Structural Business Statistics survey (SBS).

– A sampling frame should not be used as a register population because the frame will have coverage errors that can be avoided by better use of administrative sources.

9.2.2   Sampling paradigm versus register paradigm

Many lessons can be learned from the example in Section 9.2.1. The predecessor to the Structural Business Statistics survey (SBS) was a census of manufacturing establishments based on questionnaires collected by Statistics Sweden. A new administrative register then became available – the register with all income declarations for enterprises. This implies a change from a traditional survey with data collection to a register survey.

Calendar year population – a new concept

The calendar year register with all enterprises that were active during at least some part of a calendar year is not suitable as sampling frame. From the start of the year in question, the creation of the register takes about two years until all administrative data exist. The Business Register at Statistics Sweden was also based on registration data, not on activity data with tax reports and payments. With that kind of data only, creating a calendar year register was difficult. The Business Register had only been used to create sampling frames.

Coverage errors

Groves (1987) made his opinion very clear: 'Coverage error is the forgotten child among the family of errors to which surveys are subject.' We found similar attitudes when we started our work with the project described in Section 9.2.1. Perhaps we can say that coverage error was the neglected child at that time. A common idea was that 'our registers are so good that coverage error is a small problem'. This attitude was supported by measures of overall coverage errors that were only about 1–2%. Some persons also argued that 'overcoverage is about 2%, but undercoverage is also about 2%, so these errors cancel'. However, the charts in Section 9.2.1 show that these attitudes are not supported by facts.

Nonresponse or overcoverage?

Chart 9.11 describes the overcoverage in the SBS. The SBS 2004 was based on a frame population, not a register population. The sampling frame that was created during November 2004 was used as frame population for all business surveys regarding 2004, and questionnaires were sent out in January 2005. The belief was that if the SBS also used the November frame, consistency would improve.

   This procedure gave rise to the coverage errors described in Charts 9.8–9.11. The overcoverage was interpreted as nonresponse and the estimates were adjusted accordingly with imputations. The imputed values are in the columns named 'overcoverage' in Chart 9.11. This illustrates that a sampling paradigm is not appropriate when we work with register surveys.

CHAPTER 10

Building the System – The Business Register

The main purpose of all business surveys is to provide aggregate data to the National Accounts. The National Accounts Survey produces estimates of the parameters that constitute the national accounting system based on estimates from the business surveys together with other surveys such as the Labour Force Survey. The business surveys can be censuses, sample surveys or register surveys that describe different parts of the non-financial, financial and public sectors.

10.1 The Business Register and the National Accounts

When we discuss methods for creating and maintaining the statistical Business Register, everything stated here has direct implications for the National Accounts. This is highlighted in Chart 10.1, which has been used by managers at Statistics Sweden to explain the importance of the Business Register and why all surveys must be coordinated to support the National Accounts.

However, the coordination has not always been successful. For example, frame populations created before data collection have coverage errors that generate inconsistencies and register surveys have register populations that are not consistent with the sample surveys.

Chart 10.1   The system of economic statistics

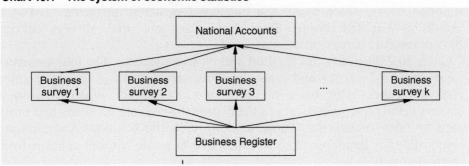

Register-based Statistics: Registers and the National Statistical System, Third Edition. Anders Wallgren and Britt Wallgren.
© 2022 John Wiley & Sons Ltd. Published 2022 by John Wiley & Sons Ltd.

Chart 1.13 is repeated here, where some inconsistencies are illustrated for relevant surveys for the employment part of the National Accounts.

Chart 1.13   Employees by industry, November 2004, thousands

Industry	Business Register Enterprises	Business Register Establishments	Employment Register	Labour Force Survey	Error margin
	(1)	(2)	(3)	(4)	(5)
Agriculture, forestry, fishing	35	37	37	26	5
Mining, quarrying, manufacturing	688	636	717	640	23
Electricity, gas, and water	21	22	28	29	5
Real estate, business activities	457	524	457	470	20
...	...	...	...	...	...
Education	382	408	431	462	20
Health and social work	836	684	675	675	24
Other service activities	142	163	175	168	13
Unknown activity	0	0	38	4	
Total	**3 763**	**3 763**	**3 924**	**3 778**	43

In this chart, there are four estimates of employment in seven industries on each row. Seven parameters have been estimated with three register surveys and one sample survey. The estimates are conflicting, to say the least. The differences between the estimates in Chart 1.13 are caused by inconsistencies between the Population, Activity and Business Registers. In addition, the variable industry is treated in a way that generates inconsistences.

How does the National Accounts use business statistics?
Chart 1.13 illustrates the dilemma faced by those responsible for the National Accounts: how do you choose between sources that give conflicting estimates for the same parameter? The problem is usually 'solved' by choosing the best source for each parameter. The source that is considered to be the best on agriculture is used for the first estimate, the best source for manufacturing is used for the second estimate, etc. This suboptimisation method leads to inconsistent subpopulations, which means the Business Register is not being used as it should be used. The Business Register was used in all these surveys, but this did not result in consistent estimates.

Chart 10.2 illustrates this method and its consequences. The four estimates for sectors 1 + 2 + 6, 3, 4 and 7 must be adjusted by 'smearing' the unexplained residual on these estimates so that they sum to 1 042 011. When all business surveys deliver their macrodata to the National Accounts, all errors and problems are also transferred. Those working with the National Accounts are responsible for handling all these problems, as a rule without support from methodologists.

Chart 10.2   Inconsistent sources with wage sums in the National Accounts
Wage sums according to the definitions of the National Accounts, SEK millions

Wage sums 2004		
Total sum for the whole economy according to the best administrative source		**1 042 011**
Wage sums from different sources:	*Sector*	*Wage sum*
Enterprises and sole traders	1 + 2 + 6	697 361
Central government	3	69 826
Local government, municipalities	4	231 897
Non-profit organisations	7	24 034
Sum of these sources:		*1 023 118*
Unexplained residual = 1 042 011 – 1 023 118 =		18 893

Those who work with the National Accounts are aware that the estimates they receive from the economic surveys have systematic errors. A common approach is to disregard the level of the estimates from business surveys and only use the relative change between the last estimate and the previous one. This way of using microdata from business surveys is not an optimal use of available surveys.

10.2   The base register for economic statistics

The statistical Business Register should be the base register for all business surveys used for economic statistics and the National Accounts. As a base register, it should have the following characteristics:

– *Standardised statistical units and identity numbers*:
  The Business Register should define important statistical units and include identity numbers that serve as links between different kinds of statistical units in the Business Register and related registers.

– *Standardised populations*:
  The Business Register should be the master sampling frame used for all economic sample surveys. In addition, the unit working with the Business Register should be responsible for creating the standardised populations used for the yearly National Accounts.

– *Standardised variables*:
  The register unit working with the Business Register should be responsible for creating and maintaining the basic classification variables used for economic statistics:

  • Industry
  • Institutional sector
  • Geographical codes for establishments

– *Main quality problems*:
  Coverage errors and errors in the classification variable for industry will generate systematic errors and inconsistencies in all economic statistics, including the National Accounts. The staff responsible for the Business Register must give priority to these two issues.

– *Main attitude problem*:
As a base register, it should be used and supported by everyone who works with economic statistics. Information on coverage errors should be reported to the Business Register, and missing values regarding sector and industry should be handled by the Business Register. Register-based statistics requires teamwork. The staff at the National Statistical Office (NSO) must understand this, and different teams should have clearly defined responsibilities.

If you hear the person responsible for the hotel survey say, 'I don't use the Business Register, my hotel register is much better than the hotel part of the Business Register,' then you have an example of this problematic attitude.

Identity numbers and access to business data

Business data in some countries are considered very confidential, and the NSO does not have access to microdata with identity numbers. The economic variables may also be rounded – then the NSO has only a very crude picture of the economy. Moreover, different taxation systems may use different identity numbers. Similar problems related with the Population Register are discussed in Chapter 2. The solution is also similar – cooperation and data sharing must be established between the NSO and the administrative authorities, including the Central Bank.

To protect confidentiality, the NSO should replace official business identity numbers with anonymous identity numbers in a similar manner as for persons. This will give sufficient protection except for large companies. Names of persons related with enterprises should also be protected.

If there are different systems of business identity numbers, then a unified system used by all authorities must be developed.

A decentralised or centralised statistical system

Countries can organise economic statistics in different ways. In some countries the National Statistical Office is responsible for the National Accounts; in other countries the Central Bank is responsible. Ministries and national institutes can also be responsible for different areas of the economy, such as agriculture, energy production, etc. These authorities produce their own statistics.

A decentralised organisation of the national statistical system can be devastating for the quality of economic statistics. Close cooperation and data sharing are necessary to overcome these problems. The goal is to have only one statistical business register used by all authorities. Problems related to the variable *industry* and a decentralised statistical system are discussed in Chapter 11.

10.3 The scope of the register and choice of object types

The scope of the statistical Business Register can be more or less ambitious. Some countries have a register describing only the private sector. To include all sectors means to include all employers, all producers, etc. in the same register. Some countries include farms in the Business Register, but they also have a Farm Register with better quality managed by the Ministry of Agriculture. A better choice would be to include the Farm Register in the Business Register.

The first priority for new register countries could be to develop a Business Register with only two kinds of units – the administrative unit *enterprise* (legal unit) and the statistical unit *establishment* (local unit) – and to attain good coverage of all sectors, both private and public.

There will be two register versions: one with all enterprises and the second with all establishments. The main part of the available administrative data concerns enterprises or legal units. This register version is absolutely necessary for register-based economic statistics. The enterprise version is created with registration data and data from taxation systems. The establishment version is created by the NSO; this is discussed in Section 10.3.1. The establishment version is absolutely necessary for regional employment statistics and regional economic statistics. More kinds of statistical units can be created, and more versions of the Business Register can be developed; this is discussed in Section 10.3.2.

10.3.1   The register with legal units and local units
This version of the Business Register has only two kinds of units:

- Legal units or enterprises. We use the abbreviation LegU. These units are administrative units created according to the rules for company registration and tax registration. These units and their identity numbers are used in all public administration.

- Local units or establishments. We use the abbreviation LocU. These units are statistical units created by the NSO for statistical purposes. They are physical units located at specific addresses where job activities take place. A *one-to-many* relation exists between legal units and local units: one legal unit can be linked to one or many local units. Since local units are physical units, they can be used for economic censuses – interviewers can go to an address and collect data.

Statistics Sweden's Business Register consists of about 1.1 million legal units, of which about 630 000 are sole traders.

We will also need an establishment version if we want to produce regional employment statistics and regional economic statistics. Statistics Sweden's Business Register consists of about 1.2 million local units or establishments.

Statistic Sweden is responsible for creating and maintaining this register version. First, 1.1 million local units are created, one for each LegU. Extensive work is then carried out at Statistics Sweden to collect information from legal units with activities at more than one local unit in order to create a register of all local units. About 8 000 questionnaires are sent out in a yearly *register maintenance survey* to all legal units that may have more than one local unit. The survey's objective is to achieve high quality in the register of local units. This survey finds new establishments and improves the variable *industry*.

Chart 10.3   Legal units by institutional sector and industry

Institutional sector: Industry:	Non-financial enterprises	Financial enterprises	Govern-ment	Sole traders	Non-profit organisations
Agriculture, forestry, fishing	11 354	0	13	236 467	546
Manufacturing, mining, energy	33 743	1	13	23 717	139
Construction	44 611	0	0	49 161	62
Trade and transport	96 626	1	5	61 606	246
Hotels and restaurants	18 598	2	0	10 966	255
Information, communication	29 010	1	1	25 807	318
Financial intermediation	10 852	2 060	10	683	1116
Real estate, business activities	157 163	15	49	112 719	10 914
Government	70	0	298	61	247
Education	8 738	0	120	14 277	985
Health and social work	14 196	0	256	17 847	979
Personal and cultural services	21 837	1	94	80 281	25 949
Total number of legal units:	*446 798*	*2 081*	*859*	*633 592*	*1 125 086*

Chart 10.3 describes a calendar year version of the register with legal units. The institutional sector is essential for the National Accounts, and industry code is used in all economic statistics. Thus, the quality of these variables is very important and the coding of these variables should be checked continuously with detailed tables similar to Chart 10.3. Some combinations in the table look odd and should be checked. For instance, the estimates in the cells with grey shade should be checked. If necessary, *register maintenance surveys* using questionnaires or telephone interviews should be carried out for groups of legal units that have suspicious combinations.

All sectors are included in Statistics Sweden's Business Register. Legal units in the public sector may consist of many establishments with many employees. As a consequence, the name Business Register is partly misleading as all public authorities and non-profit organisations are included.

According to NACE or ISIC, industry can appear in different versions. One version is the code that has been defined by the tax authorities or the Companies Registration Office. A second version is the activity code defined by the statistical office, which can be based on better information from sample surveys or register maintenance surveys. Out of about 1.1 million legal units, Statistics Sweden has improved the ISIC code for about 3 000 units with information from sample surveys and other sources. Chart 10.4 shows the different codes for some of these legal units.

Chart 10.4   Industry improved

LegU	Administrative ISIC	Statistical ISIC
BIN1	65120	53100
BIN2	70100	10822
BIN3	46360	10822
BIN4	70100	68320
BIN5	70100	24200
BIN6	70100	41200
BIN7	64920	29102

Chart 10.5   Many economic activities

LocU	ISIC	Rank	Per cent
BIN11	43120	1	60
BIN11	68201	2	20
BIN11	68100	3	10
BIN11	02101	4	5
BIN11	68203	5	5
BIN12	85323	1	51
BIN12	88910	2	49

In many cases, the statistical office has information on more than one industry for some enterprises. All economic activities should then be stored in a database together with weights that measure the relative sizes of each activity.

Chart 10.5 shows ISIC codes for two local units. Out of about 1.2 million local units, about 325 000 have more than one industry. The object type in the database table in Chart 10.5 consists of *combinations* of local units and ISIC codes. These combinations will be used to define estimators in Chapter 11.

10.3.2 The register with enterprise units and kind of activity units

With the aim of solving some quality problems, the system with two register versions in Section 10.3.1 can be expanded with register versions for more kinds of units. We have added four kinds of statistical units in Chart 10.6.

Kind of activity units (KAU) and local kind of *activity units* (LKAU) are object types created within the Business Register. There are no administrative registers with data regarding these kinds of units. These units are meant for statistical data collection. The object type enterprise unit (EU) is also created within the Business Register, but registration data can be used for this work.

Chart 10.6 Different types of units in the Business Register

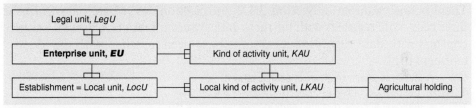

The first problem: Better estimates for industries

The idea is that KAU and LKAU should be defined so that these units can be classified as belonging to only one industry. Only a small number of these units have been created for large enterprises.

If we want to produce *economic statistics by industry*, we must create a register version with kind of activity units (KAU). The variable *industry* should be grouped according to the requirements of the yearly National Accounts. About 170 KAUs have been created at Statistics Sweden by splitting enterprise units (EU) into homogenous parts by industry.

If we want *regional statistics by industry*, we need a register version with local kind of activity units (LKAU). About 45 establishments, LocUs, have been split into 120 LKAUs by industry at Statistics Sweden. The Farm Register should be an integrated part of the Business Register and should consist of the LKAUs with ISIC = 01. These units correspond to *holdings* in agricultural statistics.

The second problem: More relevant enterprise units

The legal units have poor longitudinal quality. When an enterprise changes legal status from, for instance, *Trading partnership* into *Limited company*, then the identity number is changed. *Enterprise units* were developed to facilitate the development of demographic statistics using the Business Register. The idea is that these units should be stable over time. The legal unit that changed legal status from trading partnership to limited company could then be the same *enterprise unit* (EU) before and after the change with the same identity number EUid.

Another problem is that the legal units are not suitable for companies that have chosen to register several legal units. Such groups of legal units can be combined into an enterprise unit. In Statistics Sweden's Business Register, enterprise units have been created with the following results:

Chart 10.7   Legal units and enterprise units

Number of legal units per enterprise unit	Number of enterprise units	Number of legal units
1	944 941	944 941
2–9	27	107
10–29	10	177
30–99	5	301
100–318	3	567
	45	1 152

Out of about 1 million enterprise units, about 945 000 consist of only one legal unit. In these cases, the administrative legal units are used for statistical purposes just as they are. In the Swedish Business Register, only about 1 200 legal units are combined into about 50 statistical enterprise units.

Administrative units can be bad statistical units

There is one example in Section 3.1.2 and Section 4.6, where data from the three economic registers with business data are compared. In Chart 10.8a three registers with data regarding legal units have been matched with the identity numbers BIN.

BIN = Business identity number of each legal unit
SBS = Turnover according to Statistics Sweden's questionnaire
YIT = Turnover according to the yearly income tax returns
VAT = Turnover according to 12 monthly VAT returns

Chart 10.8   Yearly turnover for the same legal units in three sources, USD million

10.8a Incomplete composite enterprise units				10.8b Complete composite enterprise units			
BIN	SBS	YIT	VAT	EU id BIN	SBS	YIT	VAT
160006	2 301	0	0	EU1 160006	2 301	0	0
160007	2 211	0	2 239	EU1 160666		2 352	2 297
160008	1 316	1 316	0				
				EU2 160007	2 211	0	2 239
				EU2 160777		2 203	0

10.8c Final register with three enterprise units							
Id number	SBS	YIT	VAT	EU3 160008	1 316	1 316	0
EU1	2 301	2 352	2 297	EU3 160888		0	1 328
EU2	2 211	2 203	2 239				
EU3	1 316	1 316	1 328				

The problem here is that the administrative legal units in Chart 10.8a are not suitable for statistical purposes. Large companies often use many legal units when they organise their activities. To obtain meaningful economic information the legal units that belong to the same business group should be combined into *composite enterprise units* as in Chart 10.8b and Chart 10.8c. There are administrative sources describing enterprise groups that should be used for this purpose.

Units have changed

In the following example, the Annual Pay Register is compared with the Quarterly Pay Register. These registers are used by the yearly and quarterly National Accounts for estimating wages and salaries by institutional sector and economic activity. The Quarterly Pay Register is based on monthly reports from all employers delivered about two weeks after the end of each month. The Annual Pay Register is based on reports from all employers delivered about one month after the end of the year.

If we match the monthly sources and the yearly source with the matching key BIN, we find patterns that are illustrated in Chart 10.9. Again, the problem is not the record linkage as the legal units have been combined correctly. The problem is that the units have changed – there has been one takeover during the year; the legal unit BIN1 has taken over the legal unit BIN2.

Chart 10.9   Comparing gross yearly pay in quarterly and annual registers

BIN	ISIC	Quarterly	Annual	
BIN1	29	25	110	The legal units BIN1 and BIN2 have merged into BIN1
BIN2	25	84	0	

The legal units that merge have in many cases different ISIC codes. This fact will generate inconsistencies between the quarterly National Accounts that use the Quarterly Pay Register and the yearly National Accounts that use the Annual Pay Register. The legal units BIN1 and BIN2 should be combined into one enterprise unit with first ISIC = 25 (weight 84/(25 + 84)) and second ISIC = 29 (weight 25/(25 + 84)).

What is really a population?

Charts 10.10a and 10.10b compares turnover from two administrative sources for a number of legal units. Many enterprise units consist of a number of legal units. These enterprise units organise their work with accounting and tax reporting as they want.

Chart 10.10a   Each administrative system has its own object set – the effect on microdata

LegU-id	Turnover, SEK million	
	VAT	Income declaration
LegU1	31	2 161
LegU2	870	2 990
LegU3	2 084	0
LegU4	2 043	0
LegU5	0	2 040
LegU6	2 036	0
LegU7	3 998	5 939
LegU8	0	1 934
LegU9	2 558	642

The large differences between the two turnover values for these legal units should not be interpreted as measurement errors.

When we see a zero value in one source and many millions in the other source, this should also not be interpreted as a missing value in the first source. We should not impute values in such cases. Instead we should understand that this is typical for economic data from administrative units.

Chart 10.10b Each administrative system has its own object set - this gives different combinations

Income declaration	VAT-report	Number of legal units in each category	Turnover SEK million for each category	
			Income declaration	VAT-report
1. Turnover = 0	Turnover not 0	110 826	0	674
2. Turnover not 0	Turnover not 0	696 205	6 486	6 240
3. Turnover not 0	Turnover = 0	116 770	242	0

The enterprises have decided to report turnover on income declarations via 110 826 + 696 205 legal units and turnover on VAT declarations via 696 205 + 116 770 legal units. These two object sets are partially different. Some legal units (LegU1, LegU2, LegU7 and LegU9 in Chart) 10.10a) have also used both tax systems for reporting different parts of the turnover.

Total turnover is almost the same in the two sources; they differ by only 2.7%. However, if we estimate turnover by main economic activity, the estimates from the two sources will differ because the turnover has been reported via different legal units that have different codes for the main economic activity.

There are two solutions to this problem. Either the legal units are aggregated into composite enterprise units, or economic activity is treated as a multivalued variable.

From the summary table in Chart 10.10b, we can conclude that there are many more than 100 000 legal units that must be combined into composite enterprise units. This will require much work. To treat economic activity as a multivalued variable means that not only the main economic activity will be used; instead, all economic activities will be used in the estimation process together with weights.

This problem is discussed in detail in Chapter 11. Instead of working with one observation for each legal unit, these observations will be replaced by *derived units* that are called *combination objects*. How the weights are used for estimation is explained in Chapter 11.

LocU-id	1st ISIC
LocU123	62020

LocU-id	ISIC	Rank	Weight
LocU123	62020	1	70
LocU123	46470	2	15
LocU123	74103	3	15

One local unit with first ISIC 62020 is replaced by three derived units with one ISIC code for each derived unit.

Chart 10.10 illustrates that the VAT system and the yearly income declarations have different populations of legal units. Almost the same 'population' of turnover/revenue is allocated between different sets of legal units.

The unit problems we see in Charts 10.8 and 10.10 are not handled systematically. One person produces statistics with the VAT source; another person produces statistics with the income declarations. When they produce estimates by industry, the estimates will be inconsistent since the legal units in the two populations can differ regarding industry, ISIC. The inconsistent estimates are transferred to the National Accounts.

Problems and opportunities related to units in the Business Register

The *legal units* give us the opportunity to use all available *administrative data* related to business enterprises and employers. *Good coverage* of the business population is also possible by using available information in the administrative systems for registration and tax reporting. In developed countries like Sweden, where everyone must produce tax returns, these systems have very good coverage of the population of legal units we want in the Business Register. In developing countries, an informal sector can exist that is not covered by taxation systems. For such countries, we recommend that administrative registers for the formal sector be combined with area samples of households that cover both the formal and informal sectors. Avoiding double-counting requires requesting identity numbers in the household surveys and the use of administrative registers from social security systems of employees in the formal sector.

Administrative data with income statements are used to maintain Statistics Sweden's register with *establishments*. Every year, about 8 000 enterprises that have reported holding more than one establishment receive a questionnaire in order to update and maintain the quality of the establishment register. The quality of the establishment register is also improved by careful editing of income statements. For each employee, the geographical codes for the person's home are compared with the codes for the person's place of work according to the income statement. Enterprises with groups of employees with strange patterns are contacted. However, Statistics Sweden suspects that some undercoverage still exists in the establishment register.

Creating the 40 *establishment units*, 170 *kind of activity units* and 120 *local kind of activity units* noted above requires work by Statistics Sweden's profiling group. The registers with these kinds of units are expensive to maintain, and economic statistics based on these units require data collection and a substantial response burden for the companies involved. The quality problems that these units should solve are only partially solved – there are many more EUs, KAUs and LKAUs hidden in the Business Register that generate inconsistencies. Chapter 11 discusses the possibilities of creating a Business Register with *derived* KAUs and LKAUs to improve the quality of estimates by industry.

10.4 Inventory of sources

The National Statistical Office must use administrative registers to create and maintain the statistical Business Register. Here we provide a short description of the Swedish administrative systems that are used for the statistical Business Register. Similar systems are used in many countries.

Once a business has been registered with the Swedish Companies Registration Office, the business must be registered with the Swedish Tax Agency for taxation purposes. If a business is liable for value added tax (VAT), it is required to register for VAT and deliver monthly or quarterly VAT reports and if a business has employees, it is required to register as an employer for social security and Pay As You Earn (PAYE) and deliver monthly payroll reports and income statements.

An income statement is an income report for individuals issued by anyone who pays, e.g. salary, pensions, interests, or dividends. Income statements play an essential role in the Swedish society and in the Swedish tax system as they are the very basis on which the decision of the income tax that each citizen should pay is made.

Employers operating in more than one establishment should specify establishment numbers in the income statements. The data on establishment identity numbers are retrieved for Statistics Sweden and is used for statistics on employment and payrolls.

Source: Information from the Swedish Tax Board

The system with income statements is the only administrative source in Sweden that can be used to support a statistical register with *establishments*. New legislation was required so that employers operating in more than one establishment specify establishment numbers in the income statements for each employee. Without this legislation, it would not be possible to produce regional employment statistics or to maintain the establishment part of the Business Register.

All systems for tax registration and tax payments generate data for *legal units*. The systems for *tax registration* and the systems for *tax reports, tax assessment and payments* should be used for the statistical Business Register. The coverage will differ: registration means that a legal unit *may* be active, a tax return with non-zero values means that a legal unit *has* been active. In addition, different taxation systems overlap but have different coverage: units reporting VAT is one category, units reporting PAYE is another, and large corporations can organise tax reporting so that some units report for many units in a group of legal units within the same corporation.

Three conclusions regarding the sources

The first conclusion is that in a census-based system for economic statistics, the establishment (or local unit) is the main unit; but in a register-based system, the legal unit (or enterprise) is the main unit. The transition towards a register-based system will lead to the use of new registers for the updating of the Business Register with legal units and changes in the updating methods.

The second conclusion is that there are two kinds of administrative data: *registration data* and *activity data*. In Section 7.1.2, we make the same distinction for sources that can be used for the statistical Population Register. Chart 9.8 illustrates that these two kinds of sources have different coverage and that registration data may have low quality. These sources must be combined to obtain a statistical Business Register that can be used for both sample surveys and register surveys.

From Chart 9.8:

Registration data was used for the November frame for 2004:

93 114 legal units are overcoverage; they are registered as *'active'* according to the November frame but have no 'signs of life' in any administrative system regarding the calendar year 2004.

Activity data was used for the Calendar Year Register for 2004:

47 662 legal units were registered as *'never been active'*, but had reported and paid tax for 2004

88 227 legal units were registered as *'not active'*, but had reported and paid tax for 2004

171 688 legal units were missing in the Business Register, but had reported and paid tax for 2004

307 577 units was the total undercoverage in the November frame for 2004

The third conclusion is that some activity data are created after a long wait. Yearly income tax returns for enterprises will improve coverage of the Calendar Year Register; however, these are available for statistics production first during the fall the year after the income year. This means that the Calendar Year Register cannot be used as a sampling frame. Instead, it should be used as the register population for yearly economic statistics used by the yearly National Accounts.

Different kinds of information

In Chart 10.11 we distinguish between two categories of information. The first category includes sources that can improve the coverage of the Business Register. In the Population Register, both overcoverage and undercoverage result in bad quality population estimates. However, the undercoverage in the Business Register is mainly responsible for economic statistics of bad quality – overcoverage of legal units with zero production, turnover, payroll, etc. do not disturb estimates of, for example, total production by industry.

Chart 10.11   Sources with information for the Business Register

	Registration data	Companies Registration Office and Tax Board
Coverage	Activity data	Tax reports and payments from Tax Board
	Internet information	Staff at the Business Register
Quality of units and variables	Register maintenance surveys	Staff at the Business Register
	Business sample surveys	Staff at the NSO's economic sample surveys
	Telephone interviews	Staff at the NSO's profiling group

The second category of sources consists of sources that can improve the quality of the units in the Business Register and the quality of the variable *industry*. This variable can be old and not updated for many small and medium-sized enterprises.

10.5   Creating and maintaining the Business Register

The same kinds of variables that are used in the Population Register as *Identifying variables, Communication variables, Reference variables, Time references and Technical variables* should also be used here. *Updating* can be done in the same manner as with the Population Register by using time references. These matters are described in Section 7.2 and the same principles are valid for the Business Register. However, practical applications may differ even if the principles are the same.

– The shortest unit of time is *day* in the Population Register; but *month* is more relevant in the Business register. When a company is registered for VAT and PAYE, the shortest reference unit of time is month.

– How should the Business Register be used for an economic register survey? The *register population* will be defined by the administrative registers with the economic variables used for the survey. If this register population differs from the Business Register, the information in the register population should be used to improve the Business Register. Section 6.3 has an example that illustrates how two administrative registers, the Quarterly Pay Register and Yearly Pay Register, can be used to improve coverage of the Business Register.

   The staff at the Business Register is responsible for improving coverage and handling missing values for the standardised variables. This means that a register survey with activity data updates the Business Register; the register survey only needs the standardised variables in the Business Register.

– How should the Business Register be used for an economic sample survey? The *frame population* is created with the Business Register that has been updated with registration data from the Companies Registration Office and the Tax Board. However, since this information may have poor quality (illustrated in Chart 9.8), activity data from the Tax Board based on tax reports and payments is important to use.

   A simple rule can be to code a company as active after a non-zero tax report; and a company can be coded as inactive after 12 months with no activities. In Sweden, three monthly sources and two yearly sources with administrative activity data can be used to define companies as active or inactive during a reference period.

– Section 7.3 discusses how coverage of the Population Register can be improved if more sources are added. The same kind of efforts may also be necessary for the Business Register. Small family farms can be included with information from farmers' associations and suppliers to purchasers of crops. If the public sector should be included in the Business Register, the statistical office may be required to create a register with administrative authorities in central and local government, including schools and hospitals. The internet can be useful in this work.

- The sampling frames are created with the current stock of the Business Register at a specific point in time. The register populations in the administrative sources describe all legal units that have been active reporting for a specific period (year, quarter or month).
- The Business Register at Statistics Sweden is updated with the information regarding registration of legal units; these units may be active. The administrative sources contain data regarding units that have been active.
- The time lag is long between the frames and the administrative sources for the same periods. The first sampling frame for year t is created during November year *t-1*, the last administrative source for year t is delivered during November year *t + 1*.

Conclusions

The National Accounts survey needs consistent estimates from the economic surveys. In a statistical system with only sample surveys, it may seem that consistency can be achieved with the Business Register. But when an NSO starts to use administrative registers with economic data, it will be realised that economic data is much more complicated. Differences regarding coverage and problems regarding statistical units will become clear and will require long-term work with development of new competence and new methods.

CHAPTER 11

The Business Register – Estimation Methods

Most variables are easy to understand because they take only *one* value for each object. The age of a person is an example of a single-valued variable. This chapter discusses a more complex kind of variables, *multi-valued variables* and the methodological problems related to this category. A multi-valued variable can accept *several* values for certain objects, and *the number of values differs* among the objects.

Example: Industry of an enterprise: one enterprise can carry out activities in several industries at the same time. The number of industries differs between enterprises.

We discuss the variable *industry* – the variable in the national statistical system that probably generates most trouble of all. Multi-valued variables are used today in a way that causes serious errors, especially within economic statistics. To deal with these problems, the concept of *combination objects* is introduced together with estimation methods using weights. These methods can be used to reduce errors and inconsistencies.

We have borrowed the concepts *multi-valued variables* and *combination objects* from informatics. We have spent many years as students of statistics and have taught statistics at universities without any knowledge about multi-valued variables. We first became acquainted with this concept when we came to Statistics Sweden. When our first book on register-based statistics was discussed by Statistics Sweden's scientific council in 2004, some members of the council did not understand the concept: 'You mean vector variables?'

We refer to a paper by Hand (2018) in Section 1.5.2. Hand's first challenge concerns teaching statistics. Students only read about sampling errors, which gives a very limited view of statistics. Hand suggests that data quality issues should be taught, and we agree, after 25 years at Statistics Sweden. There are many important quality issues that could spark the students' interest in statistical science. One quality issue would be related to multi-valued variables. This

Register-based Statistics: Registers and the National Statistical System, Third Edition. Anders Wallgren and Britt Wallgren.
© 2022 John Wiley & Sons Ltd. Published 2022 by John Wiley & Sons Ltd.

issue is missing completely in academic statistics – a fact that has contributed to methodologists ignoring it. The praxis is to only use *first industry*, which is standard procedure in all countries.

11.1   Multi-valued variables

Chart 11.1 illustrates the problem discussed in this chapter. A user of official statistics wants to know: *'How many are engaged in health and social work?'* A producer of official statistics may answer: *'836 thousand are engaged in health and social work in the enterprise version of the Business Register with first industry of enterprise, and 684 thousand are engaged in the establishment version of the Business Register with first industry of establishment. These are estimates of two different parameters and both can be correct.'*

Chart 11.1   Employees by industry, November 2004, thousands

From Chart 1.13 **Industry** (first industry)	**Business Register**	
	Enterprises	**Establishments**
Agriculture, forestry, fishing	35	37
Mining, quarrying, manufacturing	688	636
Electricity, gas, and water	21	22
Real estate, business activities	457	524
...	...	...
Education	382	408
Health and social work	836	684
Other service activities	142	163
Total	**3 763**	**3 763**

Not many users understand this table. We use the data on agriculture to explain the peculiarities in the table.

Of 37 000 employees working on agricultural establishments, 35 000 are linked to an agricultural enterprise.

But 2 000 employees are not linked to an enterprise with 'first industry = agriculture'. They are linked to an enterprise with more than one establishment and the enterprise's first industry is manufacturing or something else.

If a statistics producer gives this kind of answer, the producer does not meet customer needs and the quality component *clarity* is unacceptable. The user may answer: *'Two parameters? I want to know how many are engaged in health and social work. For me, that is one parameter.'*

Aggregation errors

Section 4.4 discusses *aggregation* and *adjoining* as methods to create derived variables. Aggregating a qualitative variable is an operation that can produce errors and inconsistencies between different registers. In Chart 11.1, the estimate 836 has an aggregation error of at least 836 - 684 = 152 thousand employees.

In Chart 11.2, the number of employees is three in Register 1 but five in Register 3. Wage sums by industry in Register 1 differ from the other registers.

Chart 11.2 (From Chart 4.2d)

Industry, number of employees and proportion of females as derived variables – by aggregation

Register 1 – Persons

Person	Sex	Wage sum	1st Industry
PIN1	M	450 000	D
PIN2	F	210 000	D
PIN3	M	270 000	A

Aggregation

Register 2 – Job activities

Job	Person	Establish-ment	Wage sum	Industry	Sex
J1	PIN1	LocU1	220 000	A	M
J2	PIN3	LocU1	180 000	A	M
J3	PIN1	LocU2	230 000	D	M
J4	PIN2	LocU2	210 000	D	F
J5	PIN3	LocU2	90 000	D	M

Aggregation

Register 3 – Establishments

Establish-ment	Industry	Wage sum	No. empl	Prop F
LocU1	A	400 000	2	0.00
LocU2	D	530 000	3	0.33

The inconsistencies in Chart 11.2 are examples of a general problem that arises when data from different registers are integrated. Errors will occur even when all variables and identities in the registers are entirely correct. They are created during the integration process. We call these errors *integration errors*, and the errors discussed in this chapter constitute one kind of integration error that we call *aggregation errors*.

If the variable is *qualitative*, aggregation errors will occur when we aggregate variable values from *many objects to one object*. The same problem arises if one object occurs several times in a register, but with different characteristics. For example, an object has changed during the reference period of the register. The problem with aggregation errors can arise for three reasons:

- objects occur several times in a register as in calendar year registers;
- many-to-one relations, for example, establishments-to-enterprises in Chart 11.1, or jobs-to-persons in Chart 11.2; and
- variables that are originally multi-valued, such as occupation of a person or industry of an enterprise. These are discussed in Sections 11.2.1 and 11.2.2.

Objects occurring several times – calendar year registers

During the year, objects that change occur several times in a calendar year register. People move or change civil status, households change, and enterprises split or merge or change their branch of industry. Some objects can change many times during the year. All these changes create multi-valued variables. The example in Chart 9.5 is continued in Chart 11.3a, where persons moved during 2013.

Chart 11.3a   Calendar year register for the population of persons during 2013

Person	Address	Municipality	From date yyyy mmdd	To date yyyy mmdd	Weight = time at the address, years
PIN1	Address 1	1	2013 0101	2013 0517	136/365 = 0.37
PIN1	Address 2	2	2013 0518	2013 1231	229/365 = 0.63
PIN2	Address 3	1	2013 0101	2013 1231	365/365 = 1.00
PIN3	Address 4	2	2013 0101	2013 0314	73/365 = 0.20
PIN3	Address 5	1	2013 0315	2013 0925	194/365 = 0.53
PIN3	Address 6	2	2013 0926	2013 1231	98/365 = 0.27
PIN4	Address 7	2	2013 0101	2013 0605	156/365 = 0.43
PIN4	Address 8	1	2013 0606	2013 1231	209/365 = 0.57

Address and municipality are *multi-valued variables* in this example. Time can be used to generate weights for each combination of person and address. The register contains *four* persons, but *eight* combinations of person and address. As all persons live the whole year, each person's weights sum to 1. The frequency distribution of persons by municipality can be estimated using these weights (Chart 11.3b).

Chart 11.3b   Average population 2013

Municipality	Absolute frequency	Relative frequency
1	0.37 + 1.00 + 0.53 + 0.57 = 2.47	62%
2	0.63 + 0.20 + 0.27 + 0.43 = 1.53	38%
Total	4.00	100%

Calendar year registers constitute an important class of registers that sometimes have difficult estimation problems.

Many-to-one relations

In the example above, *one* register has some objects appearing many times but with different variable values. This generates multi-valued variables. In other cases, *two* registers with different object types are matched. When there are many-to-one relations, multi-valued variables can be created when these registers are integrated and qualitative variables are aggregated.

If a person has many jobs, how should information about these jobs be aggregated into information about the person? The example in Chart 11.2 above can be used as an illustration. In Chart 11.4, Register 2 is sorted by PIN.

Chart 11.4   Number of employed and wage sums in different registers

Register 1 – Persons

Person	Sex	Wage sum	1st industry
PIN1	M	450 000	D
PIN2	F	210 000	D
PIN3	M	270 000	A

Aggregation

Register 2 – Job activities

Job	Person	Local unit	Wage sum	Industry	Sex
J1	PIN1	LocU1	220 000	A	M
J3	PIN1	LocU2	230 000	D	M
J4	PIN2	LocU2	210 000	D	F
J2	PIN3	LocU1	180 000	A	M
J5	PIN3	LocU2	90 000	D	M

In Register 2, industry is a single-valued variable describing a characteristic of the object type job or job activity. PIN1 and PIN3 have two jobs; both work at a local unit within industry A and at a local unit within industry D. The traditional way to create an *industry for persons* variable is to use information about only *one* job for each person – the most important job.

In this situation when some persons have one job while others have more than one, a better solution is to define the *local unit* and *industry for persons* as multi-valued variables. In Register 2 in Chart 11.4, these variables are single-valued variables for *jobs*; but in Chart 11.5, these variables have been transformed into multi-valued variables for *persons*. The wage sums for each person are used to create weights, where the weights for each person sum up to 1.

Chart 11.5   Register 1 continued – persons and weights

Person	Sex	Local unit	Wage sum	Industry	Weight
PIN1	M	LocU1	220 000	A	22/45 = 0.49
PIN1	M	LocU2	230 000	D	23/45 = 0.51
PIN2	F	LocU2	210 000	D	21/21 = 1.00
PIN3	M	LocU1	180 000	A	18/27 = 0.67
PIN3	M	LocU2	90 000	D	9/27 = 0.33

The data matrix in Chart 11.5 is used to estimate the number of employed persons by industry. The example that begins with the inconsistencies in Chart 11.2 is finalised in Chart 11.6. According to the final estimates, there are three employed persons in industry A and D in agreement with the frequencies in Chart 11.6.

Chart 11.6   Employed by industry

Industry	Number of employed
A	0.49 + 0.67 = 1.16
D	0.51 + 1.00 + 0.33 = 1.84
Total	3.00

Both examples in this section show that multi-valued variables can arise for different reasons.

The next section contains two examples of variables that are originally multi-valued, such as occupation and industry.

11.2   Estimation methods

The variable *occupation* is created in the Occupation Register, which is a register of persons. This variable is a multi-valued variable, as many people have more than one job and can therefore have several occupations. The variable *industry* is created in the Business Register; it is a multi-valued variable of great importance. Both variables are examples that are multi-valued in the original source. As noted in the previous section, multi-valued variables are also created when registers are combined or integrated in the register system.

Multi-valued variables are difficult to deal with, but they are common and important within the register system. Multi-valued variables in statistical registers are also used in censuses and sample surveys, which means that the problem will also affect these types of surveys.

These problems are usually 'solved' in a drastic way – the multi-valued variable is transformed into a single-valued variable by using only the 'most important value' for every object. If, for example, the distribution of persons by different occupations is described, the occupations that are common as secondary activities will be underestimated. A portion of the occupational information is discarded, and estimates will be biased due to aggregation errors.

We begin with a simple example with *occupation* to show the fundamental principles for the treatment of multi-valued variables. We then look at more complicated situations that occur when the principles are applied in practice for the variable *industry*.

11.2.1   Occupation in the Activity and Occupation Registers

The estimation problem in this section can be defined as follows: how should the frequency distribution of different occupations be estimated?

> *Principle 1*
> What calculations should be done? The estimation problem should always be specified before the calculations begin. This is illustrated below where three ways of defining and solving the estimation problem are compared.

The data matrix in Chart 11.7 shows occupation and occupation code (ISCO) for six persons, of whom two have more than one occupation. The object in the matrix is *job*, which is a *relational object* that is identified by the *combination* of the personal identification number PIN for each person and the legal unit identity number BIN for each enterprise. The variable *extent*, the extent of the work, is given as a percentage of full-time work. This variable is taken from the Wages and Staff Register.

Assume that the data matrix in Chart 11.7 contains all occupational activities in a small region. How should we estimate the distribution of *persons* in the different occupations? This is our first estimation problem.

Chart 11.7   Job Register with occupational data for six persons

Job id	Person	Legal unit	Occupation	ISCO	Extent
J1	PIN1	LegU1	Statistician	2211	100
J2	PIN1	LegU2	Farmer	6111	15
J3	PIN1	LegU3	Politician	1110	10
J4	PIN2	LegU4	Hospital orderly	5132	30
J5	PIN2	LegU5	Cleaner	9122	20
J6	PIN3	LegU6	Shop assistant	5221	10
J7	PIN4	LegU6	Shop assistant	5221	50
J8	PIN5	LegU6	Shop assistant	5221	20
J9	PIN6	LegU6	Shop assistant	5221	100
Total					

The traditional approach is that each person has only *one* occupation – their *principal occupation*. This means that information on those with several occupations is discarded (grey rows in Chart 11.7); only the occupation with the largest extent of work is included. We then have a new data matrix (Chart 11.8) in which the object is *person* and where the distribution by occupation is obtained by summing the number of persons in each occupation.

Chart 11.8   Traditional register on persons with occupational information

Person	Legal unit	Principal occupation	ISCO	Extent	Weight alternative 1
PIN1	LegU1	Statistician	2211	100	1
PIN2	LegU4	Hospital orderly	5132	30	1
PIN3	LegU6	Shop assistant	5221	10	1
PIN4	LegU6	Shop assistant	5221	50	1
PIN5	LegU6	Shop assistant	5221	20	1
PIN6	LegU6	Shop assistant	5221	100	1
Total					**6**

Chart 11.9 shows the estimated occupational distribution. According to the tradition within statistics on persons, every person has the same weighting regardless of whether they work 100% or 10% of a full-time job.

Chart 11.9   Employed persons by occupation, traditional alternative 1

Main occupation	ISCO	Number	Per cent
Statistician	2211	1	16.7
Hospital orderly	5132	1	16.7
Shop assistant	5221	4	66.7
Total		**6**	**100.0**

This example shows that occupations that are common secondary occupations, such as *politician* and *farmer* (often undertaken alongside the principal occupation), are underestimated. Estimates for multi-valued variables can instead be made in a way that avoids discarding any information. This is possible if estimates are based on a data matrix with *combination objects*.

Principle 2

The basic principle is to create a data matrix so that *every combination of object and value of the multi-valued variable corresponds to one object* in the new data matrix. Objects, or rows, in such data matrices are called *combination objects*.

The data matrix in Chart 11.10 has been formed in this way. The six persons in the register of persons in Chart 11.8 give rise to nine combination objects.

Chart 11.10   Register of combination objects: person • occupation

Combination object	Person	Occupation	Extent	Weight alternative 1	Weight alternative 2
1	PIN1	Statistician	100	1	0.80
2	PIN1	Farmer	15	0	0.12
3	PIN1	Politician	10	0	0.08
4	PIN2	Hospital orderly	30	1	0.60
5	PIN2	Cleaner	20	0	0.40
6	PIN3	Shop assistant	10	1	1.00
7	PIN4	Shop assistant	50	1	1.00
8	PIN5	Shop assistant	20	1	1.00
9	PIN6	Shop assistant	100	1	1.00
Total				**6**	**6.00**

The rows in the matrix consist of all combinations of *person • occupation*. For example, person PIN1, who has three occupations, appears in three rows in the matrix. The weights according to alternative 2 have been calculated with the *extent* variable so that $0.80 = 100/(100 + 15 + 10)$, etc. The weights for each person sum to 1 in both alternatives 1 and 2. All the weights in both alternatives sum to 6, i.e. the total number of persons. Chart 11.10 illustrates several general principles:

Principle 3
The sum of the weights for *one person* (the object type that the estimation problem refers to and that was the starting point when forming the combination objects) should always be 1.

Principle 4
It follows from Principle 3 that the sum of all the weights is the same as the total number of objects (the object type that the estimation problem refers to).

In Chart 11.11, the frequency distribution of persons by occupation is calculated with weights according to Alternatives 1 and 2 in Chart 11.10.

Chart 11.11   Employed persons by occupation according to two alternatives

Occupation	ISCO	Alternative 1		Alternative 2	
		No.	Per cent	No.	Per cent
Politician	1110	0.00	0.0	0.08	1.3
Statistician	2211	1.00	16.7	0.80	13.3
Hospital orderly	5132	1.00	16.7	0.60	10.0
Shop assistant	5221	4.00	66.7	4.00	66.7
Farmer	6111	0.00	0.0	0.12	2.0
Cleaner	9122	0.00	0.0	0.40	6.7
Total		**6.00**	**100.0**	**6.00**	**100.0**

The number of employed persons (No.) by occupation is estimated by summing the weights for each occupation. *Weight alternative 1* in Chart 11.10 is summed in Alternative 1 and *weight alternative 2* is summed in Alternative 2. Aggregation errors occur because some of the occupations have weight 0 in Alternative 1, corresponding to the traditional method of calculation. Estimates according to Alternative 1 are distorted, since the frequency of certain occupations is overestimated while the frequency of others is underestimated. However, estimates according to Alternative 2 utilise all the information on occupations in the multi-valued variable.

The weights in Alternative 2 utilise the *extent* variable. This variable is found in the Swedish Wages and Staff Register. For those job positions that are not included in this register, weights must be formed from information in another register. The Income Statement Register contains *annual gross wages* for all jobs and can therefore always be used. Weights calculated from annual gross wages are somewhat different from weights calculated from *extent*. When working to create good estimates, choices must be made between different *weight-generating variables*. The choice of a relevant and functional variable is important.

The weights in Alternatives 1 and 2 are based on *extent*, but they could also be based on other variables. The weights actually used can differ more or less from the ideal weights. For certain persons, the weights for one occupation may be too large; while for others, the weights for the same occupation may be too small. The errors can partly be balanced out when forming the overall distribution of occupations. The relevant quality measurement could be a measurement of how close the estimated distribution is to the distribution that would be calculated with ideal weights.

Principle 5

Use the best weights possible even though they are not ideal.

The estimation problem in Alternatives 1 and 2 involves describing the distribution of *persons* by occupation. A third alternative, *Alternative 3*, distributes the *extent of work* by occupation. Extent or volume of work could be described by the amount of occupational activity recalculated as full-time employed persons. This calculation method is common in economic statistics, where volumes are usually measured instead of persons.

Person PIN1 has three occupations, one is full-time and the other two correspond to 15% and 10% of a full-time employed position, respectively. The matrix with the six persons represents 3.55 full-time employed positions or *full-time equivalents*. The object in the matrix in Chart 11.12 is *job*, and the variable occupation is a single-valued variable – every job corresponds to only one occupation.

Chart 11.12   Register on jobs of persons with occupational data

Person	Occupation	Weight alt 1	Weight alt 2	Weight alt 3
PIN1	Statistician	1	0.80	1.00
PIN1	Farmer	0	0.12	0.15
PIN1	Politician	0	0.08	0.10
PIN2	Hospital orderly	1	0.60	0.30
PIN2	Cleaner	0	0.40	0.20
PIN3	Shop assistant	1	1.00	0.10
PIN4	Shop assistant	1	1.00	0.50
PIN5	Shop assistant	1	1.00	0.20
PIN6	Shop assistant	1	1.00	1.00
Total		**6**	**6.00**	**3.55**

The distribution of full-time equivalents by occupation is given by summing the variable *weight alt 3* (= extent/100 in Chart 11.10) for the different occupations.

Chart 11.13   Persons and full-time employed by occupation, three alternatives

Occupation	ISCO	Alternative 1		Alternative 2		Alternative 3	
		No.	Per cent	No.	Per cent	No.	Per cent
Politician	1110	0.00	0.0	0.08	1.3	0.10	2.8
Statistician	2211	1.00	16.7	0.80	13.3	1.00	28.2
Hospital orderly	5132	1.00	16.7	0.60	10.0	0.30	8.5
Shop assistant	5221	4.00	66.7	4.00	66.7	1.80	50.7
Farmer	6111	0.00	0.0	0.12	2.0	0.15	4.2
Cleaner	9122	0.00	0.0	0.40	6.7	0.20	5.6
Total		**6.00**	**100.0**	**6.00**	**100.0**	**3.55**	**100.0**

Alternatives 1 and 2 in Chart 11.13 relate to the same estimation problem, *persons* distributed by occupation; however, they are based on different estimation methods that use different weights. Alternative 3 relates to another estimation problem, *extent of work* distributed by occupation. The focus here is on the volume of work, not on persons.

11.2.2   Industrial classification in the Business Register

Industrial classification or *industry* is another important multi-valued variable. It is created in the Business Register and is used in many registers in the register system. Again, the common practice here is to select the 'most important industrial classification' and discard information on other industrial classifications of the establishments or enterprises to which the statistics refer. This leads to aggregation errors and inconsistencies in economic statistics.

The Business Register at Statistics Sweden contains information on all branches of industry in which an enterprise is involved. It also contains details of the proportion of business carried out within each industry. Industrial classification code and the share within each industry are of good quality regarding manufacturing enterprises. The method of choosing the most important industry can cause problems when reporting industry statistics as well as time series problems. Take the example where 51% of the activities in a large enterprise in year 1 fall within a particular industrial classification, but only 49% fall within the same industrial classification in year 2. This small change can cause significant level shifts in many time series. All employees working at this large enterprise seemingly change industrial classification from year 1 to year 2.

These problems can become especially serious within regional statistics, where one local unit may be predominant. This means that a change in industrial classification will cause time series level shifts in regional series.

Slight changes are even more problematic as they are more difficult to detect, and in many cases will be misinterpreted as real changes in the economy. These quality problems can be avoided with the methodological approach to occupation presented in the previous section.

Chart 11.14 shows the industrial classification and number of employees for four local units. *The estimation problem relates to estimating the number of employees by industrial classification.* The information used by the traditional method is shown in the yellow table cells; the information in the grey cells is available but is not used.

Chart 11.14a   Business Register year 1: Data matrix for establishments (local units)

Local unit	Industry 1	%	Industry 2	%	Industry 3	%	No. of employees
LocU1	DJ	100					218
LocU2	DH	51	DJ	49			293
LocU3	DJ	40	DH	30	DK	30	156
LocU4	DJ	40	DH	30	DK	30	190

Chart 11.14b   Business Register year 2: Data matrix for local units

Local unit	Industry 1	%	Industry 2	%	Industry 3	%	No. of employees
LocU1	DJ	100					221
LocU2	DJ	52	DH	48			314
LocU3	DJ	36	DH	34	DK	30	143
LocU4	DH	45	DJ	30	DK	25	200

Chart 11.14c   Number of employees by industry, traditional estimates

Industry	Year 1	Year 2
DH	293	200
DJ	564	678
DK	0	0
Total	857	878

The number of employees is estimated by principal industry (Industry 1), which is the most common way of presenting time series based on industrial classification from the Business Register.

This leads to abrupt changes in the time series in Chart 11.14c. The industry DK does not seem to exist.

The percentages in Charts 11.14a and 11.14b show the share of each industry, which is a measure of the size of every industry at every local unit. The size measurement can be based on turnover, number of employees, or something else. We assume here that the percentages are based on number of employees.

According to *Principle 2* in the previous section, a new data matrix (see Chart 11.15) is created containing combination objects, so that every combination of objects and values for the multi-valued variable corresponds to one row in the new data matrix.

Every row in this data matrix is a combination of local unit (establishment) and industrial classification. Instead of a matrix with *four* rows referring to four local units, we obtain a new matrix with *nine* rows referring to all combinations of *local unit • industry* for each year. We can then estimate the number of employees with formula (1) for each industry:

$$Y = \sum_{i=1}^{R} w_i y_i \tag{1}$$

In Chart 11.15, $w_i y_i$ has been calculated for every row.

Chart 11.15   Data matrix with combination objects: local unit • industry

Year 1					Year 2				
Local unit	Industry	Weight w_i	No. empl. y_i	$w_i y_i$	Local unit	Industry	Weight w_i	No. empl. y_i	$w_i y_i$
LocU1	DJ	1.00	218	218.00	LocU1	DJ	1.00	221	221.00
LocU2	DH	0.51	293	149.43	LocU2	DH	0.48	314	150.72
LocU2	DJ	0.49	293	143.57	LocU2	DJ	0.52	314	163.28
LocU3	DJ	0.40	156	62.40	LocU3	DJ	0.36	143	51.48
LocU3	DH	0.30	156	46.80	LocU3	DH	0.34	143	48.62
LocU3	DK	0.30	156	46.80	LocU3	DK	0.30	143	42.90
LocU4	DJ	0.40	190	76.00	LocU4	DJ	0.45	200	90.00
LocU4	DH	0.30	190	57.00	LocU4	DH	0.30	200	60.00
LocU4	DK	0.30	190	57.00	LocU4	DK	0.25	200	50.00
Total		4.00		857	Total		4.00		878

The weights w_i sum to 4, as we are still referring to the four local units. The sums of the products $w_i y_i$ will be the same totals as before.

> The total number of employees is a given total that should not be changed when introducing weights. The weights will only affect how the employees are distributed between different industries. This is an example of a generally applicable principle.

Chart 11.16   Number of employees by industry, estimated with combination objects

Industry	Year 1	Year 2
DH	253.2	259.3
DJ	500.0	525.8
DK	103.8	92.9
Total	857.0	878.0

The time series in Chart 11.16 have been calculated with the weights w_i. We show decimals here, but estimates should be rounded in a real publication. The series here are of higher quality than those in Chart 11.14c, with relevant changes and no abrupt level shifts. The industry DK is now visible.

11.2.3   Estimates from different register versions

The Business Register has different versions for different types of units, e.g. enterprises (legal units) and establishments (local units). When several register versions contain the same multi-valued variable, the estimates from different register version should be consistent. This means that the first two columns in Chart 11.1 with employees by industry should be equal – the user that wants to know the number of employees engaged in health and social work will obtain only one answer.

The condition for consistency between these register versions is that estimates are made with weights and combination objects so that *all* information from the multi-valued variable *industry* is included in the estimates. If only the most important value of *industry* is used instead of all values with weights, the estimates will contain errors. *These aggregation errors will differ for the different register versions, so the estimates will be inconsistent.* This is shown in the example below.

Integration of data from local units to enterprises
We continue the example in the previous section with a Local Unit Register with four local units and an Enterprise Register with two enterprise units. In Chart 11.17, all information regarding year 2 in Charts 11.14 and 11.15 are shown. Information on how the local units are linked with the enterprise units is included.

Chart 11.17   Four different registers belonging to the Business Register

Register 1:	Business Register – Local units		
Local unit	1st ISIC	Employees	Legal unit
LocU1	DJ	221	BIN1
LocU2	DJ	314	BIN1
LocU3	DJ	143	BIN2
LocU4	DH	200	BIN2

The units in register 1 consist of establishments (local units)
The units in Register 2 consist of combination objects: local unit • industry

Register 2: Local units, all economic activities			
Local unit	ISIC	Rank	Weight
LocU1	DJ	1	1.00
LocU2	DJ	1	0.52
LocU2	DH	2	0.48
LocU3	DJ	1	0.36
LocU3	DH	2	0.34
LocU3	DK	3	0.30
LocU4	DJ	1	0.45
LocU4	DH	2	0.30
LocU4	DK	3	0.25

Register 3:	Business Register - Enterprises	
Legal unit	1st ISIC	Employees
BIN1	DJ	535
BIN2	DJ	343

The units in register 3 consist of enterprises (legal units). The units in Register 4 consist of combination objects: legal unit • industry

Register 4:	Enterprises, all economic activities		
Legal unit	1st ISIC	Rank	Weight
BIN1	DJ	1	**0.718**
BIN1	DH	2	0.282
BIN2	DJ	1	**0.412**
BIN2	DH	2	0.317
BIN2	DK	3	0.271

A Business Register with legal units and local units should have these four registers that are included in Chart 11.17. Register 2 is created with information from company and tax registration for enterprises with only one local unit. Information regarding enterprises with more than one local unit should be collected by the NSO. Register 4 was created by aggregating the weights in Register 2. For example, the weight 0.718 for ISIC = DJ was computed with the number of employees:

$$(221+0.52\times314)/535=0.718$$

The weight 0.412 for ISIC = DJ for BIN2 was computed in a similar manner:

$$(0.36\times143+0.45\times200)/343=0.412$$

Chart 11.18 contains four estimates for number of employees by industry. Estimate 1 with 1st ISIC for legal units is the worst, estimate 2 with 1st ISIC for local units is better, but estimates 3 and 4 are the best estimates that use all information on industry. Estimates 3 and 4 are the same except for rounding errors.

Chart 11.18   Number of employees by industry

	1. Legal units		2. Local units		3. Legal units		4. Local units
1ˢᵗ ISIC	Traditional estimate	1ˢᵗ ISIC	Traditional estimate	ISIC	Combination objects	ISIC	Combination objects
DJ	878	DH	200	DH	260	DH	259
		DJ	678	DJ	525	DJ	526
				DK	93	DK	93
	878		878		878		878

11.3   Application of the method

The Business Register must have the information described in Chart 11.17 in order to apply the methods in this chapter and produce statistics by industry, ISIC. The Business Register should include the following four registers:

1. The Business Register with establishments (local units).
2. A register with all economic activities for all local units with weights.
3. The Business Register with enterprises (legal units).
4. A register with all economic activities for all legal units with weights.

Chart 11.1 compares different estimates of employees by industry. The first two columns are repeated in columns (1) and (2) in Chart 11.19. They describe employees by industry based on data from the Business Register that have the same structure as the registers in Chart 11.17. We have used Statistics Sweden's Business Register and applied the methods described in this chapter to create the microdata regarding combination objects. The estimates based on these combination objects are shown in column (3) in Chart 11.19 together with the aggregation errors of the traditional estimates based on the main activity only.

Chart 11.19 Employees by industry November 2004, thousands

Industry	Legal units 1st ISIC **(1)**	Local units 1st ISIC **(2)**	Combination objects **(3)**	Legal–Comb. I(1)-(3)I	Local–Comb. I(2)-(3)I
Agriculture and forestry, fishing	35	37	38	2	1
Mining, quarrying, manufacturing	688	636	621	66	15
Electricity, gas and water	21	22	22	1	0
Construction	197	209	207	11	2
Wholesale and retail trade	456	453	468	12	15
Hotels and restaurants	89	93	95	6	1
Transport, communication	240	242	241	1	0
Financial intermediation	83	77	78	5	1
Real estate, business activities	457	524	526	69	2
Government	139	215	223	84	8
Education	382	408	404	22	5
Health and social work	836	684	674	162	10
Other service activities	142	163	166	24	3
Total	3 763	3 763	3 763	466 (12%)	62 (2%)

In all, 466 000 persons were allocated to the wrong activity as measured by 1st ISIC of the employer's enterprise (legal unit). The error is 12% of the total number of employees. This kind of error increases if activity is measured at the 3-, 4- or 5-digit level.	2-digit level 14% 2% 3-digit level 18% 3% 4-digit level 25% 7% 5-digit level 38% 11%

The common practice of using only the main industry and disregarding information on other activities is the main cause behind the inconsistencies in Chart 11.19. If combination object estimation is used instead, the errors that generate these inconsistencies will be reduced. The calculations in Chart 11.19 are based on real data from Statistics Sweden's Business Register.

A user of official statistics wants to know: *'How many are engaged in health and social work?'* A producer of official statistics may now answer: *'674 thousand are engaged in health and social work. That is the best estimate I can give you.'*

Conclusion
Chart 11.19 clearly shows that statistics based on the main industry of *legal units* are subject to serious errors. As most administrative sources of economic data consist of data regarding legal units, the use of combination object estimation is crucial when these sources are utilised for the production of statistics.

11.3.1 Change of industry and time series quality
Section 11.2.2 discusses the multi-valued *industrial classification* variable. The traditional methodology means that all activities in an enterprise with several industrial classifications are assigned to the largest industrial classification. This leads to level shifts in time series when the largest industrial classification for an enterprise changes. The size of the time series disturbance depends on how significant the enterprise is within the relevant industry or region.

The example below is based on reality, but the data are adapted slightly. During years 1 through 3, enterprise X Ltd. has carried out activities within several industries, but around 60% of turnover relates to industry R. During year 4, X Ltd. bought another enterprise with activities in another industry. The change of ownership took place from quarter 4 of year 4 onwards.

Column (2) in Chart 11.20 shows turnover in SEK million for all enterprises in industry R, excluding X Ltd. Columns (3) and (4) contain the total turnover for X Ltd. Column (5) has been taken from the Business Register and shows the share of X Ltd.'s activities that is carried out in industry R.

Using traditional estimation, the whole enterprise's turnover is allocated to industry R during years 1–4. Note that industry R remains as the principal industrial classification during the whole of year 4, as changes are only made at the turn of the year with the traditional methodology. Beginning year 5, none of the enterprise's turnover is allocated to industry R. By summing columns (2) and (7), we obtain the time series in column (8) that contains a time series level shift.

Chart 11.20 Estimation of turnover within one industry using two methods

Yr Q	Industry R excluding X Ltd.	X Ltd. before pur-chase	X Ltd. after pur-chase	X Ltd.'s share within Industry R	Weight with traditional estimation	X Ltd.'s contribution to industry R, traditional estimation	Industry R traditional estimation	X Ltd.'s contribution to R with combina-tion objects	Industry R with combi-nation objects
(1)	(2)	(3)	(4)	(5)	(6)	(7)	(8)	(9)	(10)
1 1	7 684	7 354		0.60	1	7 354	15 038	4 412	12 096
1 2	7 086	7 086		0.60	1	7 086	14 172	4 252	11 338
1 3	8 142	6 788		0.60	1	6 788	14 930	4 073	12 215
1 4	9 853	8 387		0.60	1	8 387	18 240	5 032	14 885
...	...	...		...	...	...	...	...	...
4 1	13 071	9 259		0.57	1	9 259	22 330	5 278	18 349
4 2	13 127	9 509		0.57	1	9 509	22 636	5 420	18 547
4 3	11 253	9 499		0.57	1	9 499	20 752	5 414	16 668
4 4	12 921		15 881	0.21	1	15 881	28 802	3 335	16 256
5 1	12 782		12 397	0.21	0	0	12 782	2 603	15 385
5 2	13 360		12 634	0.21	0	0	13 360	2 653	16 013
5 3	11 098		11 621	0.21	0	0	11 098	2 440	13 538
5 4	12 888		13 209	0.21	0	0	12 888	2 774	15 662

In accordance with the estimation methodology based on combination objects, the shares in column (5) should be used as weights. By multiplying the enterprise's turnover in columns (3) and (4) with the weights from column (5), we obtain the part of the turnover that relates to industry R. By adding this part of the turnover to column (9), which shows the turnover for the other enterprises within industry R, we obtain the time series in column (10), which describes the industry's turnover without any time series level shift.

Errors can be considerably reduced by using weights, and the quality of the time series increases substantially.

The estimation methods are compared in Chart 11.21. The series in column (10) has been estimated using the weights in column (5). Even if these weights are not perfect, they are considerably better than the traditional weights in column (6).

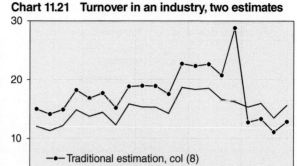

Chart 11.21 Turnover in an industry, two estimates

The traditionally estimated turnover series is affected by a time series level shift caused by abruptly changed aggregation errors. For the period year 1 to year 4, the traditional method gives an overestimate of the industry's turnover. This error becomes even more significant during the fourth quarter of year 4. From year 5 onwards, the turnover within industry R is underestimated.

Multi-valued variables – summary of recommendations
A variety of important variables in the register system are multi-valued. The current way of handling these variables produces estimates with aggregation errors. These errors can be reduced when producing estimates by using combination objects and weights. This chapter describes a series of different estimation problems with multi-valued variables, and suggestions are made for solutions to these problems. The above example of the change in industry shows how relatively simple methods, such as using weights, can bring quality improvements, even though the weights are not perfect.

11.3.2 Transformation of weights
The weights for the different industrial classifications in the Swedish Business Register are primarily based on turnover. They should be transformed when calculating estimates for other variables.

Example: When estimating number of employees, the weights based on turnover should be transformed into weights that are suitable for number of employees. This transformation can be based on a model that describes the relation between turnover and number of employees. The model is based on appropriate statistics that describe employment and turnover for industry-specific local units.

Chart 11.22 shows this transformation for local unit LocU3. The weight based on turnover is multiplied by the number of employees per turnover. These values must then be divided by a constant so that the total for every local unit is 1.

Chart 11.22   Transformation of weights

Register of local units			Aggregated data		Register of local units		
Year 1		Weights	Models for different industries		Transformed weights based on model		
Local unit	Industry	based on turnover	Employees Turnover SEK m		adapted for estimation of number of employees		
LocU3	DJ	0.4	DJ	0.5	LocU3	DJ	$\dfrac{0.4\cdot0.5}{0.4\cdot0.5+0.3\cdot0.6+0.3\cdot0.7}=\mathbf{0.34}$
LocU3	DH	0.3	DH	0.6	LocU3	DH	$\dfrac{0.3\cdot0.6}{0.4\cdot0.5+0.3\cdot0.6+0.3\cdot0.7}=\mathbf{0.30}$
LocU3	DK	0.3	DK	0.7	LocU3	DK	$\dfrac{0.3\cdot0.7}{0.4\cdot0.5+0.3\cdot0.6+0.3\cdot0.7}=\mathbf{0.36}$

According to Principle 3, the sum of the weights for *one local unit* (the object type that the estimation problem refers to) should always be 1.

The transformed weight for the share of the local unit belonging to the capital-intensive steel industry DJ is lower than the original weight, which was based on turnover (0.34 compared with 0.4). The example also shows that when using turnover, the most important industrial classification is DJ; but when using number of employees, it is presumably DK.

This again shows that the principle of only using the 'most important' value of a multi-valued variable can cause problems. Furthermore, when registers contain multi-valued variables, weights can be calculated that are adapted for different estimation problems.

11.4   A decentralised or centralised statistical system?

Section 10.2 notes that a decentralised organisation of the national statistical system can be devastating for the quality of economic statistics. All economic surveys used by the National Accounts should be consistent. Close cooperation and data sharing are necessary to achieve this kind of coordination. In this section, we discuss the following topics:

- How can the calendar year version of the Business Register be used to coordinate and improve economic statistics and the National Accounts?
- How can the calendar year version of the Business Register be used when revised estimates from business sample surveys are produced?
- How can the present method of choosing the best source for the National Accounts estimates be improved?

The calendar year register in Chart 11.23 was created at the end of 2008. This register gives a rather accurate picture of the target population of the National Accounts for 2007. These accounts are based on several business surveys, most of them done by Statistics Sweden, but some were done by other Swedish statistical institutes. The problem of coordination and consistency is a difficult problem, as the Swedish economy is not described by a single survey with consistent estimates. Instead, it is described in a piecemeal fashion with many surveys for different parts of the economy. This is in sharp contrast to social statistics where this piecemeal approach is not used. For instance, the Labour Force Survey describes the whole labour market with consistent estimates.

The Structural Business Statistics survey (SBS) gives consistent estimates for non-financial enterprises and sole traders, but the National Accounts prefers to use better sources for some subpopulations such as agriculture, energy and real estate. In addition, the supply–use tables for about 400 products created by the National Accounts indicate that the National Accounts is based on a very large number of sources, which means that the coordination issue is a serious problem.

Assume that we want to estimate Sweden's GDP for 2007. If we compute value-added for all units in Chart 11.23, then we want *all* units to have been included in our GDP estimate. No units should be overlooked, and no units should be included more than once. If we compare the frame populations and the register populations for all surveys with the calendar year population, we will find the gaps and overlaps in the system of surveys used by the yearly National Accounts. We have made such comparisons and found many gaps and overlaps.

11.4.1   The Calendar Year Register and the National Accounts
One way of using the calendar year version of the Business Register is to start to work with comparisons of frame and register populations of surveys used by the National Accounts. This will be a learning process that will take many years. The aim of the work is to coordinate future surveys so that gaps and overlaps are gradually reduced.

Chart 11.23   Calendar year register for 2007, legal units by sector and industry

Sector: Industry:	Non-financial enterprises	Financial enterprises	Central and local government	Sole traders	Non-profit organisa-tions	Total
Agriculture, forestry, fishing	11 354	0	13	236 467	546	**248 380**
Manufacturing, mining, energy	33 743	1	13	23 717	139	**57 613**
Construction	44 611	0	0	49 161	62	**93 834**
Trade and transport	96 626	1	5	61 606	246	**158 484**
Hotels and restaurants	18 598	2	0	10 966	255	**29 821**
Information, communication	29 010	1	1	25 807	318	**55 137**
Financial intermediation	10 852	2 060	10	683	1 116	**14 721**
Real estate, business activities	157 163	15	49	112 719	10 914	**280 860**
Government	70	0	298	61	247	**676**
Education	8 738	0	120	14 277	985	**24 120**
Health and social work	14 196	0	256	17 847	979	**33 278**
Personal, cultural services	21 837	1	94	80 281	25 949	**128 162**
Total	**446 798**	**2 081**	**859**	**633 592**	**41 756**	**1 125 086**

Our vision is that economic statistics could be structured in a way that is similar to the structure of register-based social statistics. Chart 1.14 shows some output from the register-based census. All registers with census-like information are based on the same population, the population on December 31; and all registers, such as the Employment, Education and Income registers, are perfectly consistent.

In a similar way, all economic registers could be based on the same population, the population in the calendar year register; and all registers used by the yearly National Accounts should be made consistent.

A second way of using the calendar register is to revise the estimates of existing business surveys. The coverage errors in all sample surveys describing 2007 can be reduced if the frames that were created earlier are compared with the final calendar year register. After calibration of sampling weights, new revised estimates can be used for the work with the yearly National Accounts.

11.4.2   Choosing the best source for the National Accounts

The Swedish National Accounts currently uses macrodata from many surveys. When two surveys give estimates for the same parameters, the source considered best is used. Agriculture is an example where the National Accounts uses the Farm Structure Survey (FSS) from the Board of Agriculture instead of the Structural Business Statistics (SBS) survey from Statistics Sweden. The SBS uses 1st ISIC = 01 to define agriculture, but the FSS includes only the agricultural part of all establishments above defined thresholds. This means that the FSS uses data for *holdings*. The different kinds of units in the Business Register are explained in Section 10.3.2.

These two surveys have populations that are very inconsistent. Selander (2008) describes the undercoverage in the Business Register regarding agricultural enterprises. Statistics Sweden (2007b) describes the corresponding overcoverage. In Chart 11.24, the population in the Farm Register (FR) has been matched against the units in the Business Register (BR).

Chart 11.24   Coverage errors in the Business Register (BR) compared with the Farm Register

Data from 2004 and 2005	Number of units in		Activity	Coverage	
	FR	BR	In ISIC 01	errors in SBS	SEK millions
1. ISIC 01 in BR but unit not active		6 787	Not active at all	Overcoverage production	1 110 1.8%
2. ISIC 01 in BR but unit not in FR		64 498	Active, but not in 01		
3. Unit in both FR and BR	67 112	67 112	Active, and in 01		
4. Unit in FR but not in BR	8 696		Active, and in 01	Undercoverage turnover	4 425 6.8%
Total number of units:	75 808	138 397			

Source: Based on Table 2.2 in Wallgren and Wallgren (2010)

The Farm Register (FR) contains 75 808 units, and the Business Register (BR) contains 138 397 units coded as active in ISIC 01.

Out of these 138 397 units, 6 787 (line 1) were not found in any administrative source for the reference year. Therefore, they are overcoverage in the Business Register. In the SBS, these inactive enterprises were treated as nonresponse and positive values were imputed. This gave rise to an overcoverage error of 1 110 million SEK, or 1.8% of the total production value for ISIC 01.

Furthermore, 64 498 units in the Business Register (line 2) were coded as active within ISIC 01. However, since they did not belong to the Farm Register, we suspect that the ISIC classifications of these enterprises in the Business Register are wrong in most cases. They are not active in ISIC 01, but in another branch of industry. These enterprises are then overcoverage regarding ISIC 01 and undercoverage regarding other branches of industry.

On line 3, there are 67 112 units that are active in ISIC 01 in both the Farm Register and the Business Register. We consider these units as correctly classified.

Moreover, 8 696 units in the Farm Register were not found in the Business Register (line 4). We can classify these as undercoverage in the BR. Due to this undercoverage, the estimate of turnover in SBS was 4 425 million SEK too low, or 6.8% of the total turnover value for ISIC 01.

Conclusions of the present method

The SBS cannot be used for the agricultural part of the National Accounts because the coverage errors are unacceptable. Instead of using the 138 397 units with SBS data, the National Accounts uses data regarding the 75 808 units in the Farm Structure Survey (FSS). But there are some adverse consequences of this strategy: the 64 498 units on line 2 in Chart 11.24 are active and produce in other industries. These units will be excluded from the National Accounts and total GDP will be underestimated.

In addition, the 67 112 units on line 3 probably do not have production only in industry 01. Nonetheless, other kinds of production for these units will be excluded from the GDP estimate with the present method.

The new method

The currently used method results in biases and inconsistencies due to gaps and overlaps of population units. Instead, we are proposing a new method based on *integrated microdata*. In this case, microdata from the SBS and the FSS should be combined. This requires cooperation and data sharing. The combined data set should be subsequently used to produce suitable estimates for the National Accounts. In Chart 11.25 we outline the present estimation method, and in Chart 11.26 the new method is illustrated.

The charts illustrate that the gaps and overlaps in Chart 11.25 are eliminated in Chart 11.26 where the two surveys SBS and FSS are combined. In Chart 11.25 the current method is described. The largest enterprise LegU3 has been split into two kind of activity units (KAU31 with activity only in ISIC = 20 and KAU32 with activity only in ISIC = 24) and data for these have been collected with questionnaires. The smaller enterprises are used as they are, and 1st

ISIC is used when estimates are computed. This illustrates the method that is used in the SBS.

In Chart 11.26, all combinations of legal unit and industry are listed in database table 2. These combination objects are derived KAUs and instead of sending out questionaries, value added is estimated with the industry weights *w* for each KAU.

Chart 11.25   Statistics Sweden's present way of using business surveys

Business Register

1. Legal units database table

LegU id	1st ISIC
1	01
2	16
3	20
4	24
5	01
6	02

3. KAU database table

KAU id	1st ISIC
11	01
21	16
31	20
32	24
41	24
51	01
61	02

LegU 3 has been split into KAUs: 31,32

Statistics Sweden's Structural Business Statistics survey (SBS)

LegU id	KAU id	1st ISIC	Value added
1	11	01	423
2	21	16	375
3	31	20	6 600
3	32	24	6 184
4	41	24	3 456
5	51	01	1 476
6	61	02	984
		Total	19 498

Excluded by National Accounts

2. Industry database table

Combination objects = LegU • Industry

LegU id	ISIC	Rank	Per cent
1	01	1	50
1	02	2	30
1	10	3	20
2	16	1	60
2	02	2	40
3	20	1	51
3	24	2	49
4	24	1	90
4	10	2	10
5	01	1	60
5	02	2	40
6	02	1	55
6	01	2	45

Board of Agriculture
Farm Structure Survey, FSS

LegU id	ISIC	Value added within ISIC 01
1	01	282
5	01	750
6	01	500
Total		1 532

Counted twice by National Accounts

Estimates from the FSS

ISIC	Value added
01	1 532
Total	**1 532**

Estimates from the SBS

ISIC	Value added
01	1 899
02	984
16	375
20	6 600
24	9 640
Total	**19 498**

Estimates by the National Accounts

ISIC	Value added
01	1 532
02	984
16	375
20	6 600
24	9 640
Total	**19 131**

Chart 11.26   A new estimation method based on integrated microdata

Business Register

1. Legal units database table

LegU idno
1
2
3
4
5
6

2. Industry database table

Combination objects = LegU • Industry

(Derived KAU database table)

(KAU idno)	LegU idno	ISIC	Per cent
11	1	01	67
12	1	02	20
13	1	10	13
21	2	16	60
22	2	02	40
31	3	20	51
32	3	24	49
41	4	24	90
42	4	10	10
51	5	01	51
52	5	02	49
61	6	02	49
62	6	01	51

Register survey of legal units

3. SBS and FSS integrated microdata

LegU idno	ISIC	Industry w	Value added	VA • w
1	01	1.00	282	282
1	02	1.00	85	85
1	10	1.00	56	56
2	16	0.60	375	225
2	02	0.40	375	150
3	20	0.51	12 784	6 520
3	24	0.49	12 784	6 264
4	24	0.90	3 456	3 110
4	10	0.10	3 456	346
5	01	1.00	750	750
5	02	1.00	726	726
6	02	1.00	484	484
6	01	1.00	500	500
			Total	19 498

From FSS

Board of Agriculture

Farm Structure Survey, FSS

LegU idno	ISIC	Value added within ISIC 01
1	01	282
5	01	750
6	01	500
Total		1 532

revised weights

Estimates by the National Accounts

ISIC	Value added
01	1 532
02	1 445
10	402
16	225
20	6 520
24	9 375
Total	**19 498**

Estimates from the combined SBS and FSS in database table 3 can be used directly by the Natinal Accounts

When the National Accounts replaces the agricultural part of the SBS in Chart 11.25, legal units 1 and 5 are partially excluded. Instead of 423 + 1476 only 282 + 750 are included in the GDP estimate. The agricultural part of legal unit 6 is counted twice. It is included in the estimate 1 532 and in the estimate of value added of forestry (ISIC = 02), 984. The estimate of total GDP is biased downwards due to excluded units, and biased upwards due to counting some units twice. Observing and understanding the errors is not possible because the National Accounts has only access to aggregated estimates.

Unbiased estimates can be produced when the microdata from the SBS and FSS are combined as in Chart 11.26. In a real application, the record linkage will not be as easy as in Chart 11.26. There will be some many-to-one links, for instance, husband and wife can be two sole traders in the Business Register but one holding in the Farm Register. Such problems can be handled after careful editing, where we search for unit errors.

When information from the SBS and FSS is combined, the weights for industries (the last column 'Per cent' in the Industry database table 2 in Chart 11.26) can be improved. We used these improved weights to produce the final estimates. Instead of obtaining conflicting estimates from inconsistent surveys, the National Accounts will now arrive at consistent estimates based on all available sources.

It should be noted that the combination objects in the industry database table 2 in Chart 11.26 can be interpreted as *derived* Kind of Activity Units (KAU). This kind of unit is mentioned in Section 10.3.2. These units are created by the Business Register and require data collection when used in business surveys. Only a small number of this kind of units have been created. Creating and maintaining this part of the Business Register is costly and data collection is also costly. Creating the derived KAUs discussed here does not generate costs and requires no data collection – administrative registers with economic data are already available for use.

Choose the best method instead of the 'best source'

Administrative registers are the most important sources for economic statistics in the Nordic countries, and this will also be the case in other countries. In countries like Sweden, the Structural Business Statistics survey is mainly based on tax return data from legal units. This survey and register-based payroll statistics are two very important sources for the National Accounts. Administrative registration and taxation systems with information on legal units are indispensable for creating and maintaining a Business Register with good coverage.

However, the methods used today must be improved. In this section, we propose the following:

– The calendar year register should be used as standardised population for the yearly National Accounts. All surveys used for the yearly National Accounts should be consistent with respect to the calendar year register.

– The classification variable *industry* should not be used as it is used today, when all activities of a legal unit or an establishment are assigned to 1st ISIC. Instead, *industry* should be treated as a multi-valued variable and weights should be used to estimate all activities.

– The National Accounts estimation method should be improved. Instead of using aggregated macrodata, consistent estimates based on integrated microdata from several business surveys should be developed.

11.5 Conclusions

Many important classifications in the register system are multi-value variables. These variables are also used in sample surveys, but as a rule these are imported from the register responsible for the variable.

Our impression is that multi-valued variables have been neglected by statistical science. The general practice at statistical offices is to use only the main value of the variable and disregard the other values. This practice gives rise to errors and, as these errors will be different depending on the statistical unit used, the errors will also generate inconsistencies.

Industry

Industry is an important multi-valued variable in economic statistics. How this variable is handled significantly determines the quality of economic statistics.

This chapter presents estimation methods that can reduce the problems associated with this variable. We have tested and elaborated these methods so that the methodology can be used in many applications. We cannot see any reason for the use of the traditional method that neglects information on all values except for the main value of the multi-valued variable. When an NSO publishes statistics and states, for example, that turnover within enterprises with main industry X was *x* million, we may think this is correct. But it gives a false picture of the real world. The enterprises with main industry X also produce and sell goods belonging to other branches of industry; and enterprises with other main activities also produce goods belonging to activity X.

CHAPTER 12

Censuses, Sample Surveys and Register Surveys – Conclusions

The national statistical system is based on a combination of censuses, sample surveys and register surveys. The previous chapters described how a register-based national system can be developed. The transition from a traditional system without registers to a register-based system will provide a more cost-effective system with more opportunities and improved quality.

This final chapter presents an overview and draws some conclusions regarding survey design, survey methods and survey quality. These three aspects are closely related: the survey design determines what methods should be used, and these methods determine the quality of the statistics produced. We begin the chapter with a discussion of the statistical paradigm and registers.

12.1 Attitudes towards the register-based census

'In other words, a register-based census is a rudimentary tool for grappling with the increasing complexity of modern society.' This was Eurostat's conclusion regarding Statistics Denmark's register-based census 1981. Lange (1993) describes the attitudes that met the first register-based census. There was widespread distrust of data retrieved from registers, and this was especially true regarding methods and data used for the census and the LFS.

Why were register-based statistics not accepted at that time? Lange lists several reasons:

- *Ignorance*, people from countries without national population registers could not understand the situation in the Nordic countries with their national administrative systems used frequently by citizens and authorities.
- *Bad quality* of population registers in non-Nordic countries. Many member states compared their population registers with census data and found errors in register data regarding 10–20% of the population in small areas. Statisticians from these member states had difficulties understanding that the situation in the Nordic countries was quite different.

Register-based Statistics: Registers and the National Statistical System, Third Edition. Anders Wallgren and Britt Wallgren.
© 2022 John Wiley & Sons Ltd. Published 2022 by John Wiley & Sons Ltd.

- *No understanding of the possibilities* availed by a system of linked registers. The population can be georeferenced so that all registered persons and all dwelling households are placed in time and space according to the census requirements. This is often not understood.

- *Registers are static* – they contain old information that is never changed; hence, registers cannot provide information regarding new issues. However, new issues that are important for administration will gradually be included in registers. New legislation regarding, for example, partnership or economic issues for taxation, will quickly be incorporated in the registers.

- *Registration anxiety* arose in connection with censuses in Germany, the Netherlands and Sweden during the 1980s and 1990s. Nonresponse rates in the Swedish LFS increased and a negative debate took place regarding a Swedish register for research (not kept by Statistics Sweden). Nevertheless, the central administrative registers are well known and used by the citizens in the Nordic countries. We believe the public accepts these registers and trust the register keepers. However, public opinion was judged not to accept registers in many member states. Note that the negative feelings were towards the traditional census. No negative feelings arose in the Nordic countries when the census was replaced by registers.

Denmark was the first Nordic country to join the European Union. When Statistics Denmark published the census for 1981, which was completely register-based, their statistical colleagues in the union must have been shocked – 'What is this: no interviewers, no questionnaire?' The Danish census was the first register-based census. A new survey method had been developed and the traditional census parameters had been estimated with a new methodology. The output was the same, but the methodology was completely different.

The initial response was that this could not be accepted – the traditional, well-established, and scientific census methods should be used! Such strong feelings and negative attitudes have been associated with register-based statistics from the beginning.

However, after some years, Eurostat understood that the Danish methodology was an important new contribution and became positive. Eurostat arranged a seminar with all member states, including two new member states, Finland and Sweden, which also had register-based statistical systems.

'It is quite clear that the Member States find themselves in the paradoxical situation of having to face a number of budget cutbacks at the same time as providing users with an increasing volume of high-quality relevant information.' So began Yves Franchet, then Director General for Eurostat, a seminar on the use of administrative sources for statistical purposes (Eurostat, 1997). The quote illustrated the need for more effective statistical systems.

During the seminar, some countries expressed their doubts regarding quality. Confidentiality concerns were also discussed, but the seminar was a turning point – register-based statistics were now accepted.

The discussion in the Statistical Journal of the IAOS

The following themes regarding the classical or traditional census versus the register-based census were discussed in the Statistical Journal of the IAOS (2020). The discussion started with a paper by MacDonald (2020) and our contribution to the discussion is in Wallgren and Wallgren (2020).

Population censuses: are statistical dinosaurs able to adapt?

Population and Housing Censuses: an overdue and old fashioned instrument or still a modern, severely needed and steadfast tool?

Themes to be discussed:

On the definition of a census:

Should the census be defined by its unique methodology or are the outputs the main element of its definition? An item related to the definition is the question if the criteria of individual enumeration, universality, simultaneity, defined periodicity, and capacity to produce small-area statistics, are still relevant as essential features of a census?

Source: IAOS (2020).

The term *register-based census* was used when the new methods for register-based statistics were introduced. The term indicates the great potential of administrative registers – even the costly and difficult population and housing census can be replaced with yearly statistical registers. With the methods that Statistics Denmark developed for the first register-based census 1981 it was possible to estimate the same parameters as in a traditional census (= universality). The population was placed both in time and space (= simultaneity) and consistent statistics for small geographic areas and small population categories were possible.

Nowadays, the term *register-based census* is no longer necessary. We do not need think of the whole census output – every user can use the statistical registers he or she needs.

Themes to be discussed, continued:

On the necessity of a census methodology:

Some authors question the theoretical soundness of the register-based census. Readers are invited to contribute to the discussion to give their opinion if such a theoretical base is really needed for a census, and why so?

Readers might agree with the statement that the risk with the register-based census approach is that it reverses the paradigm of statistics: it is only measuring what is available, instead of properly defining first a concept and then develop a methodology to measure it.

Source: IAOS (2020).

To say that with the register-based approach we '*only measure what is available, instead of properly defining first a concept and then develop a methodology to measure it*' is a serious misunderstanding. Decades of hard work were necessary to develop the national statistical system and the national administrative systems to make the register-based census possible. Chart 1.5 illustrates how long time it took to make the register-based census possible in the Nordic countries. The intention was to make it possible to estimate the required parameters for a traditional census. This work included the following steps with survey system redesign:

– Improving the national systems of identity numbers of persons, dwellings and enterprises.
– Improving legislation regarding citizens' obligations to report migration.
– Improving legislation regarding employers' obligations to report where each employee works.
– Improving legislation to support data sharing and cooperation between different actors in the national statistical system.
– Work with standardisation of addresses and a Dwelling Register.
– Creating a statistical Population Register and a statistical Business Register.

When we measure 'what is available', we find that the administrative systems are handling important matters of life: births, deaths, migration, education, employment, income, health care and social support. For economic statistics, we obtain payroll data, turnover, import and export data, profit and loss statements and balance sheets. All of these sources are available for statistics production.

When discussing the theoretical soundness of the register-based census, we should analyse the theoretical census model developed by Statistics Denmark (1995). Their model for a population and housing census based on registers has clear definitions, standardised procedures and a specific methodology for place and time. *They defined the concepts first and then developed the methods to measure them.*

If identity numbers (PIN) are included in the household sample surveys conducted by the NSO, then the quality of the register-based statistical estimates can be verified.

'Theoretical soundness': our interpretation of this term is that a survey (census, sample survey of register survey) *with* quality assessment complies with the scientific method and a survey *without* quality assessment does not.

Themes to be discussed, continued:

On relevance of census taking and census results:

Censuses are generally seen as very costly projects. The readers are invited to give on the discussion platform their opinion on to what extent the limited use of census results, in particular for evidence-based policy making, is worth the huge cost?

And finally, discussion contributions are expected on the statement that, in the current fast moving world, to produce census results only every ten years is not relevant any more.

Source: IAOS (2020).

Chart 12.1 compares monthly data from the Swedish register-based census with the corresponding values from a fictional traditional census every ten years. Two municipalities about 100 km apart in central Sweden are compared. The Population Register, which is a part of the register-based census, shows clearly that important changes have taken place in both municipalities. The chart also shows the importance of timely small area statistics, which can only be produced with the Population Register – the most important part of Sweden's statistical system.

Chart 12.1   Monthly population data for two Swedish municipalities 2000–2020

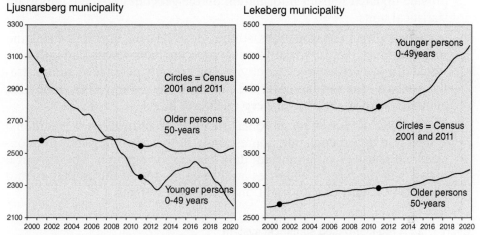

Source: Statistics Sweden's Population Register   Time series analysis by the authors

Traditional censuses cannot provide timely information about the important changes in Ljusnarsberg 2013 and 2017, or in Lekeberg 2015. Chart 12.1 also illustrates the importance of small area estimates. These municipalities are geographically close but have quite different demographic trends.

12.2   The new national statistical system

This section describes the main differences between a traditional statistical system and a register-based system. A traditional system without registers is based on area frames. The Population and Housing Census is the method that must be used to collect social statistics for small areas. A register-based system is based on the public administrative systems that continuously or periodically generate new and updated data on persons and organisations.

Chart 12.2 describes the main features of the register-based system. Some important parts of the new system are missing in a traditional system:

– The *base registers*; in a traditional system there are only area frames for area samples of households, farms and establishments.

– *Time references*; all demographic events in the base registers can be placed in time. The census day or week is the only time reference in a traditional system. The monthly population statistics in Chart 12.1 can be created using the time references in the Population Register. Traditional censuses can only provide us with a snapshot every tenth year.

- The *identity numbers* are used as links in the system. In a traditional system names, birthdates and birthplaces are used for probabilistic record linkage. The Nordic method of record linkage is based on the register system approach.

- The residences of persons and households, and the location of jobs and establishments are *geocoded statistical units*. Regional statistics and local area level statistics are made possible because the administrative systems provide updated register information on the geocodes of these important statistical units.

- *Activity registers* regarding jobs, studies, hospital care, unemployment and social support, etc. In a traditional system the interviewers can only ask: 'What did you do last week?' Activity data are only possible in administrative systems – the systems cover the entire population and all the time – the administrative systems never sleep and never forget.

- *Longitudinal registers*, longitudinal surveys are difficult in a traditional system due to accumulating nonresponse. Both activity registers and longitudinal registers are important data sources for research. The statistical registers in the Nordic countries avail research opportunities that are not possible in countries with traditional statistical systems.

Chart 12.2   A register-based statistical system

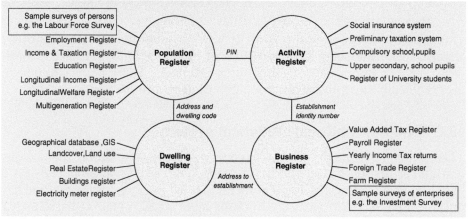

12.2.1   The system of base registers

The system in Chart 12.2 is a model of the whole national statistical system. All surveys and registers with any of the identity numbers in the system of base registers can be linked together. Social statistics, economic statistics and agricultural statistics become parts of the same statistical system. These parts can be linked at the microdata level, which makes possible data sharing, quality assessment and control of consistency.

This is not possible in a traditional system and social, economic and agricultural statistics live their own lives more or less independently and with inconsistencies that are difficult to measure. Data sharing, quality assessment and control of consistency are not possible without a national system of base registers with identity numbers.

We stress these facts – the key benefits of a register-based statistical system are due to these features of the system of base registers.

Record linkage with the register system method

When a country has developed a system of base registers as in Chart 12.2, record linkage will be very efficient and flexible. The base registers can be matched, which means that object sets that consist of different object types can be matched. For example, persons can be linked with the establishments where they work, and jobs can be linked with the dwellings where persons live.

The methods used for record linkage differ radically between countries that have a system of base registers and countries that lack such a system – high quality deterministic record linkage with identity numbers as matching keys versus probabilistic record linkage where there are errors in the matching keys. The Nordic countries had a clear strategy for this – if a country develops a system of base registers as in Chart 12.2, then the NSO can use the matching keys in the system to achieve very efficient and large-scale record linkage.

New opportunities for quality assessment

All registers in the system in Chart 12.2 can be combined with all other registers, and all sample surveys can be combined with all registers. This means that register populations, units and variables in different surveys can be compared. These comparisons should be used for a new kind of quality assessment.

Today, we always perform quality assessment for sample surveys – we always compute standard errors or confidence intervals for the estimates. We hope that NSOs in the future always will perform quality assessment of sample surveys and register surveys by comparing coverage with other registers. Coverage errors should be analysed and reduced – the register system makes this possible.

The administrative systems update the matching keys

In the Population Register, a combination of two matching keys is used to georeference the residence of persons and households: PIN and address code. In the Activity Register, a combination of two matching keys is used to georeference jobs and study activities: PIN and establishment numbers. The Population and Activity Registers also contain time references that have information on when these matching keys have been changed due to persons moving to a new dwelling or obtaining a new job.

The important issue here is that the administrative systems update these matching keys so that persons and jobs can always be placed in both time and space according to the census requirements.

Geocoded statistical units

Administrative systems provide updated information on residential addresses and addresses to establishments. The homes and jobs of persons can be linked to digital maps and the information is updated promptly.

These maps should be created from coordinates available from the National Land Survey, thus defining the geographic dimension in the register system. These maps can be maps with coordinate points. The polygon map of real estates should also be included, if it is available.

Information from maps can be linked to the Real Estate Register. Statistics on land cover and land use, pollution and other kinds of statistics based on measurements with geographic location can then be included in the register system. A paper by Haldorson (2019) discusses statistical-geospatial integration and methods for creating statistics at the local area level.

12.2.2 Activity registers and longitudinal registers

Administrative systems remember everything; the systems cover the whole population and all the time. In contrast, collecting activity data in sample surveys is difficult – people do not remember and have difficulties placing what has happened to them on the time axis.

Persons in the Population Register, and enterprises, authorities and organisations in the Business Register are all actors in society. Their activities generate large amounts of activity data within many administrative authorities that handle taxation and welfare services, such as education and medical and social care.

Object types and variables in the Activity Register

The register consists of three different sections: *gainful employment or job activities, study activities* and *other activities related to the labour market*. As previously noted, the register consists of relational objects. Every such object is identified by three matching keys: personal identification number of employees or students, organisation number and local unit number of employers or schools from the Business Register.

Data on job activities are based on the employer's annual or monthly income statements, which give the income for every combination of *employee* and *establishment*. Tax reports provide information on persons who are active as *self-employed*. In addition, there are administrative data on social insurance that employers are required to have for their employees. All of these kinds of data can be used to create a register with job activities.

Data on *study activities* are based on data from schools and universities. A variety of registers exists for different kinds of students. These registers should contain the students' personal identification numbers and details about the school or place of study, which in turn is a local unit in the Business Register. Commuting can be identified for students in the same way as commuting for gainfully employed persons. A personal identification number linked with a dwelling identity and a local unit number give the location of the place of residence and the place of work or studies.

Data on *other labour market-related activities* could also be included in the register. Different authorities have information on military service, sickness benefits, disability pensions, employment policies, registered unemployment and institutional medical care. This information can give a complementary picture of the attachment of persons to the labour market over and above information on gainful employment and studies. The administrative sources contain information that can locate activities in terms of time, even if the quality of this information is sometimes poor.

Activities – a neglected distinction?

Statistics regarding students and employment can be misinterpreted as statistics on persons. When we report the number of students by study programmes, many students are double counted when they follow more than one programme. In a similar way, data collected from employers can be reported as persons employed by industry. Here we have double counting when persons have more than one job. In these cases, we actually work with activity data. The statistical units are study and job activities, not students or employees. Statistics Sweden's Job Register contains about 8.0 million jobs but only 5.3 million persons have these jobs.

When labour market supply and demand meet, relations are created between individuals and enterprises/organisations. These relations are important for labour market statistics and are described by many statistical variables. Regarding these relations as objects is convenient. Demographic statistics that describe how, for example, the range of gainful activities changes through *job creation* and *job destruction* are relevant in the study of labour market statistics. The Activity Register should therefore contain the birth and death times of the activities.

Statistics based on the Activity Register

The Activity Register is used for new kinds of statistics. One example is given here. The population 16 years and older is classified into seven disjunct categories describing job activities with information in the Job Register:

1. *Emp-wh yr: Employed the whole year January–December*
2. *New-emp: Received a new employment during the year; the first month of this employment is in the interval February–December, the last is December*
3. *End-emp*: Ended an employment during the year, the first month is January and the last month of this employment is in the interval January–November
4. *Emp-part yr*: Employed part of the year, February–November
5. *Emp and Self*: Both employed and self-employed during the year
6. *Self-emp*: Self-employed during the year
7. *Without*: Persons without any job as employed or self-employed

The following chart from Statistics Sweden (2007a) gives an example of statistics based on the Activity Register. A complicated set of activity data has been aggregated into variables describing persons. The longitudinal table below is more advanced than traditional official statistics and is suited for more specialised users. The grey cells show persons who belonged to the same category in both years.

Chart 12.3   Transitions from 2003 to 2004, men 25–54 years. % of each category 2003

2003	2004 Emp-wh yr	New-emp	End-emp	Emp-part yr	Emp and Self	Self-emp	Without	All
Emp-wh yr	83.4	5.5	5.2	2.0	1.9	0.5	1.4	100.0
New-emp	65.0	12.2	11.5	5.8	2.5	0.4	2.7	100.0
End-emp	51.5	14.0	8.4	6.6	2.4	1.9	15.2	100.0
Emp-part yr	44.0	17.2	7.6	15.3	2.4	0.9	12.6	100.0
Emp and Self	7.5	0.8	0.7	0.4	80.7	9.2	0.6	100.0
Self-emp	3.1	0.6	0.2	0.2	8.1	84.2	3.6	100.0
Without	5.2	8.0	0.8	3.5	0.5	2.7	79.3	100.0

Source: Based on Statistics Sweden (2007a) Register-based Activity Statistics (in Swedish). Background Facts – Labour and Education Statistics 2007:2.

Creating longitudinal registers

Kardaun and Loeve (2005) describe and compare longitudinal analyses in some statistical offices. Register-based longitudinal surveys are common in Scandinavian countries, and longitudinal sample surveys are common in Anglo-Saxon countries. They also report that most longitudinal surveys are person oriented and only a few are business oriented.

Register surveys have the advantage of complete coverage of time. Thus, register surveys are suitable for longitudinal analysis. However, a prerequisite of using administrative data this way is that the longitudinal register has been created with sufficiently high longitudinal quality regarding *variables* and *statistical units*.

The administrative units in enterprise statistics are difficult to follow over time; they merge and split but can continue to use the same identity number. Persons may also change identity numbers, or sometimes two persons can have the same identity number, or an old number can be reused. Administrative data regarding changes of identity numbers should therefore be maintained by the statistical office. As soon as evidence of duplicate identity numbers is found, this should be checked and the information stored. The unit within the statistical office that receives data on persons should use this information when the national identity numbers *PIN* are replaced by anonymous identification numbers *anPIN*. This will improve the longitudinal quality of the data used for the production of statistics.

Longitudinal registers are often used for two kinds of analysis – *cohort analysis* and analysis of *transitions* between categories. Chart 12.3 is an example that shows transitions between different categories in the Job Register. Chart 12.4 is an example with cohort analysis.

Chart 12.4 compares persons with lower and higher levels of education as they try to enter the labour market. Six cohorts, consisting of all persons who completed upper secondary school in 1987–1992, are followed during the years 1988–1993. Their transition into gainful employment can be compared with the corresponding six cohorts of students graduating from university. These years are of particular interest as it was a period when the labour market changed dramatically. All persons belonging to these 12 cohorts were studied via longitudinal registers. The circles in the charts below represent the share of gainfully employed persons one year after completing their educational programmes. The curves show the development of the share of gainfully employed persons within each cohort.

Chart 12.4   Per cent employed after completing education 1987–1992

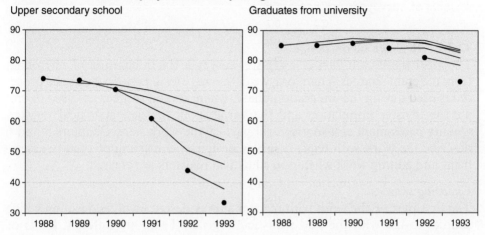

The most serious crisis in the Swedish labour market since the 1930s occurred in the early 1990s. Chart 12.4 shows how the economic downturn at the beginning of the 1990s changed the possibilities for young persons to enter the labour market. The share with gainful employment one year after completing upper secondary studies changed dramatically during these years. The development was not so depressing for those with university degrees – this appears to show that higher education studies were worthwhile.

12.3   Survey design

In a traditional statistical system, there are no identities that can be used to combine microdata from different surveys. Since most surveys are sample surveys that cannot be combined, the design of each sample survey can be considered independently of other sample surveys.

Survey design for register surveys is fundamentally different from sample survey design. When working with register surveys, we always have a number of potential sources that can be used in different ways for a number of surveys or registers. We need to think of systems of surveys and systems of registers.

12.3.1   Sample survey design

Lohr (2009) defines *sample survey design* as: 'survey design means the procedure used to select units from the population for inclusion in the sample'. We note that the concept of sample survey design is closely related to the most significant part of a probability sample survey – the selection of the sample. The purpose is to design inclusion probabilities to achieve high quality in the sense of small standard errors of the estimates. Thus, quality assessment regarding sampling errors is central in sample survey design.

Survey design within the realm of classical sampling theory deals with one survey at a time. With a sample survey, *one* set of microdata is collected, and this set of microdata will be used for *one* specific purpose. Thinking about systems of surveys is not necessary.

12.3.2 Register survey design

The most significant part of a register survey is the integration of microdata from different sources. Therefore, we define *register survey design* as the procedures used during the integration phase, where several sources are used to create the register population and the variables in the new statistical register. Quality assessment is also a central part of the work here. Chapters 5 and 6 describe the work with register survey design as consisting of quality assessment and editing work when new statistical registers are created.

Choosing sources

The first part of the work with register survey design involves deciding which sources to use with the new statistical register. There are different kinds of potential sources:

- Relevant administrative registers can be used as input.
- Existing statistical registers at the NSO:
  - At least one base register will be used as a source of standardised variables and for creating the register population in the new register.
  - Some statistical registers may be used as sources with variables.
- Existing statistical registers and sample surveys can be used for the work with quality assessment of the new register.

Different scope of the research objectives

A traditional sample survey, e.g. the Labour Force Survey (LFS), reports only one set of data – the data collected in the survey. The team responsible for the LFS may describe their mission as '*We collect LFS data. We analyse and report LFS data.*'

There is a risk that those working with register surveys also understand their mission in the same narrow manner: '*We are responsible for administrative data from source X and we analyse and report X data.*'

All the opportunities of the register system are not exploited when a team responsible for a register survey understands their mission in this narrow way. Instead, the team should use all relevant registers in the system to *analyse and describe their subject field.*

Example: The Register of University students

This register receives administrative data from all universities concerning what the students want to study, what they actually study, and the results of their efforts and examinations. The traditional approach is to report these sources only. However, there is more information in the register system about these university students:

- Do they also work to earn money, and how much money do they earn?
- From where did they come? Region? Socio-economic group?
- What did they do before their university studies?
- What happened after they finished their university studies?

If the team responsible for university statistics decides to answer these questions, instead of only reporting university studies, it can use more registers in the system.

The procedure for creating a statistical register

To conduct a register survey means to create a specific statistical register and to use that register to produce survey estimates. The newly created statistical register will also be a part of the register system.

The work with sample surveys is usually divided into different stages. The work with a register survey can be similarly divided into different phases:

1. Determining the research objectives and the inventory phase:
What statistical requirements should the register fulfil? What sources are available when creating a new register? What administrative sources are available and what existing statistical registers in the system can be used?

When planning a register survey, data are as a rule already available. Determining the research objectives and conducting an inventory and preliminary analyses of sources can then be done simultaneously.

2. Contacts with data suppliers and the receipt of administrative data:
Maintaining good contacts with the data suppliers is an essential part of the work with register surveys. Checks and editing take place of the received administrative data.

3. The integration and editing phase:
Different sources are integrated into a new statistical register to be used for producing statistics. The work can be divided into three parts:

a. How should the existing sources be integrated so that the register will contain the required population? The administrative data are checked and edited so that the population is the required one. Different sources are matched, and objects are selected. Time references are processed to create the object set for the point in time or time period in question.

b. What processing should be carried out to check and correct object definitions? Administrative data are checked and edited so that the definitions of the statistical units are those required. Derived units are formed in the new register.

c. What processing should be carried out to create the variables in question? The variables in the administrative sources are checked and edited. Steps are taken if variable values are missing. Different sources are matched; variables are selected and imported into the new register. Derived variables are formed in the new register.

4. The estimation phase:
Selection of estimation methods for carrying out calculations and compiling tables. Should weights be used? Tabulation and dissemination.

Work with a register survey should be structured so that the persons working with the survey are aware of the three simultaneous processes illustrated in Chart 12.5.

Chart 12.5   The three parallel processes with a register survey

Create and use a register for producing statistics	Quality assurance	Documentation
Creating a register 1. Determining the research objectives and the inventory phase 2. Receiving data 3. Integration phase: – create population – create statistical units – create variables **Calculations and tabulations** 4. Estimation methods for carrying out calculations and compiling tables	Contacts with data suppliers Checking on receipt of data Causes and extent of missing values Causes and extent of nonmatch Evaluating quality of register population, units and variables Questionnaires for register maintenance Investigate inconsistencies between sources Documentation of quality issues	When microdata are collected from administrative registers, metadata for these sources should also be collected. Definition of object set for every source. Definitions of imported variables. What checks and processing have been carried out in the different sources? What is known about the quality of the different sources? The new register should then be documented.

The first process consists of work to create the statistical register, and is described in points one to four above. The newly created register and the register-statistical processing should simultaneously be *quality assured* and *documented*.

Quality assurance

Quality issues are observed when working with integration in step three above, and quality improvements are made. These issues should be analysed and documented.

The quality of the register should be reviewed and described using various quality indicators. Documentation is also an important part of quality assurance. Incorrect and uncritical use of administrative data can be prevented by the existence of metadata, which give information on possible comparability problems.

Documentation

A statistical register should be available to many users at the statistical office. To enable this efficient use of data, all the registers should be documented in such a way that everyone can use and understand the documentation.

Metadata have a very significant role in the work with register-based statistics. When linking and matching different registers, knowledge of the definitions and awareness of any comparability problems are necessary.

The documentation of processing methods is also important in facilitating the development of methodology and the exchange of experiences.

12.3.3   Creating register variables

Once the new register's population has been created and edited, variable selection is carried out. Variables of statistical interest are imported from different sources. Other variables to be used for editing should also be imported. These editing variables are correlated to the register variables so that edit rules can be formulated.

- Adjoined variables and aggregate variables are formed by matching with registers containing other object types.
- The register variables are edited. Should unreasonable and missing values be replaced with imputed values?
- Derived variables should be formed using the register's variables.

The created registers will be parts of the register system, which imposes certain requirements related to coordination and cooperation. The editing and processing should be adequate for other usage in the register system. This helps to avoid duplicate work and increases the value of the created register.

Estimating variable values with a rule
A statistically meaningful variable can be formed using a number of administrative variables. In this section, we use y for the derived statistical variable and x_1, x_2, ... for the administrative variables. Rules for how the derived variable should be defined are used in the following examples. But the rules used here are *models* that are based on knowledge of the sources and the subject field. The calculations here are not precise; the y variable may contain errors even if the x variables are correct. These errors in the y variable are called *model errors*.

Example: Sex in the Austrian Census Register – a rule gives priority to sources
Lenk (2009) describes how sex in the final census register is a derived variable based on the variable for sex in seven different administrative sources. In some cases, there is conflicting information regarding sex. Then a rule must be defined that determines how sex in the census should be defined based on information in the sources.

Based on experiences of the quality of the seven sources, the sources and different combination of the sources are given different priorities in the script used in the data processing. Some complicated combinations result in a loop-back process, and the sex of such persons is determined based on the judgement of an analyst. Giving priority to sources is a common way of creating derived variables in a register.

Estimating variable values with a causal model
An alternative to using rules is to analyse the relation between the desired variable y and the administrative variables x_1, x_2, ... by building a statistical (causal) model. The derived variable can then be created using the statistical model.

There are two steps involving different data matrices when creating derived variables using a statistical model:

1. The first data matrix contains *test data* from, say, a sample survey containing both the y variable and the x variables. With this data matrix, a model is first put together to show how to best estimate y for the given values of the x variables.
2. The model is then used on the second data matrix, the *register's data matrix*, where only the x variables exist. With the estimated model, y values for all units in the register are calculated with the help of each unit's known x values.

The advantage of a statistical model compared with a rule-based model regarding knowledge of the subject field and judgements is that a good statistical model shows how to best use many administrative variables. The model can contain many variables, as opposed to a rule-based model, purely based on knowledge of the subject field and judgements. With this kind of model, a statistical variable y from the test data set is used to *calibrate the measurements* for the derived variable in the register. The test data set is usually a sample – this method of creating derived variables is thus another example of how sample surveys can be used in the work with register-based statistics.

Example: Employment status in the Employment Register from 1993
The first version of the Employment Register 1985–1992 used the following rules:

– Social insurance data for a part of the year that included November:[1] if total income per month (including sickness benefit) was higher than SEK 200, the individual was classified as gainfully employed in November.
– Social insurance data for full year: if total annual income was higher than SEK 21 800, the individual was classified as gainfully employed in November.

This rule-based methodology has serious disadvantages. Young and old persons are classified according to the same income limits even though their salary levels are different. Income patterns, the distribution between permanent and temporary jobs, the system for sickness benefits and the structure of the social insurance data have changed over time. The result of all these changes is that statistics from different years are not comparable, despite the use of the same rules.

To solve the problems of comparability outlined above, a new derived variable was introduced in the 1993 version of the Employment Register. For those persons who participated in the Labour Force Surveys (LFS) in November 1993, social insurance data were combined with employment status according to the LFS. Using regression analysis, models were built for different sex–age combinations for these test data. The different groups were given different income limits in this way, but all the limits corresponded to the employment definitions in the LFS.

---

[1] November was the month for the traditional census in Sweden.

Another advantage is that high quality administrative variables can greatly affect the classifications, while low quality variables have little effect. A broad outline of the analysis follows:

1. Regression models are estimated with test data with known employment status according to the LFS. The LFS employment status is the y variable with two categories (gainfully employed/not gainfully employed), and the social insurance data are the regression model's x variables. The test data are divided into subgroups, which have the same type of social insurance data, age categories and sex. Separate analyses are carried out for every subgroup.
2. Using the estimated model for a subgroup, an estimated y value is calculated using the x variables from the social insurance data. If the analysis succeeds, those classified as gainfully employed in the LFS will have estimated y-values that are markedly different from those not gainfully employed.
3. A cut-off value is determined so that those with estimated y values on one side of the cut-off value are classified as gainfully employed, and the remaining persons are classified as not gainfully employed. The limit is set so that the number of persons classified as gainfully employed will be the same size as the corresponding number according to the LFS in the test data.
4. These cut-off values for the different subgroups are then used so that all persons in the register population will be classified using the administrative variables in the social insurance data.

The important advantage of the new methods was comparability over time. For example, the 2001 November LFS was used to produce new income limits via new regression analyses for the 2001 version of the register. These new limits have the same definition of gainfully employed in the LFS as previously. Relevant comparisons could be carried out between different years in this way.

Qualitative derived variables that have been formed using a statistical model should be estimated so that *model errors* can be judged. In this case, the model errors are *classification errors*. With a good statistical model, both *net errors* and *gross errors* should be minimal. Chart 12.6 shows a comparison between the old and the new methods of defining employed persons in the Employment Register.

The estimated *gross error* in Chart 12.6 is an estimate of the share of incorrect classifications in the entire register, whereas the estimate of the *net error* is an estimate of the systematic error in the method of defining gainfully employed in the Employment Register (assuming that the LFS gives correct estimates).

Model errors in derived variables
When the values of derived variables are estimated with a model, the model or classification errors can be regarded as random. The model errors should be examined with existing or special sample surveys, and the results of these surveys can be used to estimate systematic and random model errors. One example of how model errors can be measured and described is presented here.

Chart 12.6   Classification errors in the old and new Employment Register 1993

Number of persons in test data	Estimate in *new* Employment Register			Estimate in *old* Employment Register		
	Employed	Not employed	Total	Employed	Not employed	Total
Employed LFS	22 360	1 158	23 518	22 472	1 046	23 518
Not employed LFS	1 068	6 872	7 940	1 329	6 611	7 940
Total	23 428	8 030	31 458	23 801	7 657	31 458

Per cent of total number of persons	Estimate in *new* Employment Register			Estimate in *old* Employment Register		
	Employed	Not employed	Total	Employed	Not employed	Total
Employed LFS	71.1	3.7	74.8	71.4	3.3	74.8
Not employed LFS	3.4	21.8	25.2	4.2	21.0	25.2
Total	74.5	25.5	100.0	75.7	24.3	100.0

Classification error	Net error: 74.5 – 74.8 = –0.3%	Net error: 75.7 – 74.8 = 0.9%
	Gross error: 3.7 + 3.4 = 7.1%	Gross error: 3.3 + 4.2 = 7.5%

We understand that estimates in sample surveys are unreliable if the number of observations is small – the same is true also for estimates in register surveys when we have random classification errors. Chart 12.7 can be used to describe the quality of the employment variable in the Employment Register at an aggregate level. We show here that the quality of individual estimates can be described with methods of statistical inference.

In the Labour Force Survey, 31 458 persons have been interviewed and classified as employed or not employed. If we trust the quality of the LFS, the probabilities of the two kinds of model or classification errors of the derived variable *employed* in the Employment Register can be estimated as 5% and 13%, respectively.

Chart 12.7   Model errors in the Employment Register 1993

Number of persons in test data	Estimate in Employment Register				Estimated classification errors:		
	Employed	Not employed	Total		Employed	Not employed	Total
Employed LFS	22 360	1 158	23 518		95%	5%	100%
Not employed LFS	1 068	6 872	7 940		13%	87%	100%
Total	23 428	8 030	31 458				

If we assume that all employed and unemployed are classified with these risks of error in the register, we can estimate quality components as follows:

The number of persons classified as employed in the Employment Register is the sum of two independent stochastic variables with binomial distributions: the number of persons classified as employed in the Employment Register among truly employed plus the number of persons classified as employed in the Employment Register among truly not employed. Chart 12.8 compares the true numbers of employed and not employed persons in two domains with the corresponding expected numbers and standard errors based on the estimated classification errors in Chart 12.7.

Chart 12.8   Quality of estimates for two domains in the Employment Register

	True number of:		No. of persons	Expected number of	Standard error of
Domain	Employed	Not employed	in domain	employed in the register	the register estimate
1	100	100	200	95 + 13 = 108	4.0
2	50	10	60	47.5 + 1.3 = 48.8	1.9

An estimated employment rate of 54% (=108/200) in domain 1 with 200 persons is expected to have a systematic error of approximately 8/200 = 4 percentage points, with a standard error of 4.0075/200 = 2 percentage points.[2]

An estimated employment rate of 81.3% (=48.8/60) in domain 2 with 60 persons is expected to have a systematic error of approximately –1.2/60 = –2 percentage points, with a standard error of 1.9/60 = 3 percentage points.

The information about the systematic error could be used to produce adjusted estimates; and the information about the standard errors could be used to indicate that table cells are based on too few observations.

The example in this subsection shows that inference methods can be developed for random errors in register-based statistics. However, these methods are primarily for the producers of register-based statistics and can be used when writing quality declarations. These inference methods should not be used to publish thousands of standard errors for all estimates; this will only confuse users. We are sceptical regarding confidence intervals for register-based estimates for two reasons:

- The users of register-based statistics often base their conclusions on simultaneous comparisons of many, sometimes hundreds, of estimates; and
- there are non-random errors that invalidate confidence intervals.

12.4   Survey quality

The statistics producer needs knowledge about quality in order to work with quality improvements and as a basis for informing users regarding the fitness of the published statistics for use.

The basic method for quality assessment of administrative sources and statistical registers is systematic comparisons with related sources and surveys. Knowledge regarding coverage issues is gained by comparing object sets and register populations. Comparing related variables from different sources for the same units provides knowledge of quality issues regarding units and variables. This has been illustrated by many previous examples in the book, and the method is used in Chapter 6 where the editing of register data is discussed. The aim of editing is to find and correct errors, and this work gives important knowledge on data quality.

---

[2] The sum of these two binomial random variables has the variance $100 \times 0.05 \times 0.95 + 100 \times 0.13$ •$0.87 = 16.06$.

12.4.1   Quality of registers and register surveys

The quality of the *register* should be described in general terms so that potential users can see whether it suits their purposes. The description should relate to the various areas of application that may be of interest. We distinguish between three ways of using registers and the corresponding quality aspects:

- *Cross-sectional quality*: what comparisons can be made between categories within the register?

- *Time series quality*: what comparisons can be made over time at the aggregate level? This is the quality level that is required for official statistics. The users of official statistics usually need time series.

- *Longitudinal quality*: what comparisons can be made at the micro level over time? This is the quality level that is required for registers that will be used by researchers. The researchers usually need longitudinal registers.

The main quality issue for many users is *comparability*. Comparability over time at the microdata level is necessary for a longitudinal survey. For individual observations in the data matrix, measuring changes over time should be meaningful.

In longitudinal sample surveys, the variables and the questionnaire can easily be kept constant over time. In longitudinal register surveys, however, administrative variables are often changed due to changes in the administrative systems.

In the example above with *Employment status in the Employment Register from 1993*, we show how the LFS can be used to improve comparability. The longitudinal Income Register at Statistics Sweden improves comparability by analysing old definitions and new income definitions.

Example: Longitudinal income

The Income Register at Statistics Sweden uses data based on income self-assessments. New variables are added and replace old variables as the taxation systems and different systems for transfers change over time.

This will be a problem in a longitudinal register, where the definition of, say, disposable income will change. This should be handled by efforts to calculate two income variables for each person. Disposable income according to both the old definition and the new definition should be included in the register.

Chart 12.9 shows data regarding the income of six persons over three years. The definition was changed in year 2.

Chart 12.9   A longitudinal income register, data for six persons

PIN	Year 1 Old def.	Year 2 Old def.	Year 2 New def.	Year 3 New def.
1	241 202	237 302	227 799	222 502
2	178 399	176 703	173 700	179 699
3	143 800	152 900	152 900	163 800
4	130 800	130 501	130 501	142 103
5	421 798	424 203	404 203	375 499
6	196 300	216 398	211 601	222 099

Disposable income, SEK per year

This kind of income data can be used in different ways. The transition between decile groups based on ...

old def year 1 – old def year 2 ...and...

new def year 2 – new def year 3 ...will give a picture of the transition between income classes.

12.4.2   The integration process – integration errors

Sampling errors have long been regarded as the most important error in sample surveys. Therefore, sampling designs and estimation methods have been developed to reduce this kind or error.

There is no sampling phase in register surveys. Instead, this kind of survey is dominated by the integration phase, where data from different sources are integrated into a new statistical register. The register population and derived objects are created during the integration phase; variables are imported from different sources and derived variables are created. The kinds of errors that have their origin in the integration phase should be called *integration errors*. This category includes coverage errors, matching errors, missing values due to non-match and aggregation errors.

Distinction should be made between three different situations with regard to the possibility of improving and describing the quality of register surveys.

1. Register surveys where we can obtain detailed measures of one or more kinds of errors and correct or reduce these errors. After the correction, we have no quality measures for the corrected estimates, as there are no more sources that can be used for comparisons.

2. Register surveys where we can obtain measures of one or more kinds of errors (perhaps from a sample survey) but not at a detailed level. Errors can therefore not be corrected or reduced, but we have quality measures on an aggregate level.

3. Register surveys where we have not been able to measure errors.

Group 1 should be maximised, and group 3 should be minimised. The method for quality assessment consists of comparisons with other sources or surveys in the production system. We need not restrict ourselves to administrative sources in this work. When we suspect that a certain category of a register population has quality flaws, we can conduct a register maintenance survey and send questionnaires to this category of units to measure and improve quality.

Existing sample surveys should also be used to evaluate the quality of administrative sources. Sometimes we can conduct a sample survey with the primary aim of evaluating registers and register surveys.

12.4.3   Frame errors

Twelve of the 13 chapters in Cochran (1963) are devoted to sampling errors. In the last chapter, Cochran mentions measurement errors and nonresponse. During the last few decades, much effort has been devoted to nonsampling errors. Measurement and nonresponse issues are today regarded as central issues in survey methodology.

Thirteen out of 14 chapters in Särndal and Lundström (2005) are devoted to nonresponse issues; frame errors are mentioned in the last chapter. There are currently no established methods for handling frame errors. We believe that this kind of nonsampling error has been overlooked and that the errors can be substantial. Development in this field is necessary, and these errors can only be reduced by register-statistical methods. If staff learn how to create registers with good coverage, then all surveys using these registers will benefit from the good coverage.

The first step is to become aware of the frame errors. At a statistical office where sample survey theory is the predominant paradigm, registers are used to produce frames and thereafter data are collected. As a rule, the quality of the frame population will never be analysed. Instead, new frames will be created, followed by a new round of data collection.

A statistical office, where those responsible for, say, a business survey want to use administrative data, may follow the same procedure – except that instead of sending questionnaires to the sampled enterprises, they use administrative data for the enterprises in the sample. Response burden and costs will decrease, but the frame errors will be the same. If administrative data are used in this restricted way, the most important quality of administrative sources has not been used – the capacity for good coverage.

If registers are used instead to create both the frames and calendar year populations, then it will be possible to become aware of the frame errors. The preliminary estimates for the sample surveys based on frames can also be revised with information from the calendar year register, and the methods used to create frames can also be improved so that frame errors become smaller.

12.5   Organising the new production system

This section discusses which organisational models are suitable for the work with register-based statistics. Enterprise architecture, data warehousing and a system-based work with missing values are discussed.

12.5.1   Enterprise architecture and the register system

Many national statistical offices are currently discussing process orientation and an enterprise architecture based on the Generic Statistical Business Process Model (GSBPM). This model is discussed in a paper by Eltinge, Biemer and Holmberg (2013).

We interpret this model as an attempt to standardise the production of primarily sample surveys. The aim of standardisation is to save costs (mainly costs for developing and maintaining IT tools) and improve quality. Chart 12.10 gives an overview of the processes in the model. The seven main processes in the chart are divided into a total of about 40–50 subprocesses. We interpret the model as a model for streamlined production of separate sample surveys. The model was introduced at Statistics Sweden when we still were working there.

We are doubtful if register-based statistics and complex surveys such as the National Accounts have benefited from the GSBPM-model. Register surveys and the National Accounts survey stress the importance of a systems approach. We interpret the GSBPM-model as being the opposite – instead of systems thinking, a fragmented approach to survey methodology is used. The survey process is divided into small more or less independent parts which can be discussed one at a time.

Chart 12.10   Different parts of the survey production process

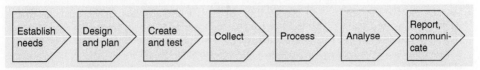

Advanced and complex statistical surveys such as the Population Register or the system of base registers cannot, of course, be described by the GSBPM-model. Even if we take the simplest possible register survey as the Quarterly Pay Register in Chart 6.5 (which is repeated below), the GSBPM-model will not help us.

'All relevant sources should be considered simultaneously' means that three sources, three statistical registers and the National Accounts survey must be considered together. Consistency editing of the three registers reveals that quarterly wage sums for insurance companies are wrong and that the Business Register suffers from undercoverage and missing values for sector and industry.

From Chart 6.5   All relevant sources should be considered simultaneously

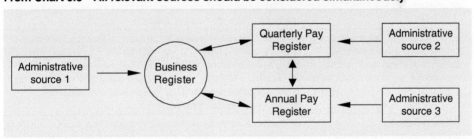

These errors were found during the work with quality assessment and consistency editing of the Annual Pay Register. These finding should result in corrections of the Quarterly Pay Register, improvement of the Business Register, and corrections of the Quarterly National Accounts that used the Quarterly Pay Register. Treating these fours surveys separately is bad practice. The GSBPM-model may lead to such bad practice and, therefore, is not suitable for complex surveys as register surveys often are.

12.5.2   The register system and data warehousing
Data warehousing as an IT methodology for storing microdata for statistical purposes is discussed at some national statistical offices. A short overview is given in Goossens (2013). We provide some comments here regarding data warehousing and the discussion of the register-based production system.

Quotations from Goossens	Our comments
1. How to make optimal use of all available data sources?	1. This is one of our main themes in our book. The optimal use of sources is attained with new register-statistical methods, not with new IT tools. The development of a system of base registers is necessary to make optimal use of all available data sources.
2. ... realising maximum reuse of available statistical data ... demands ... a metadata catalogue that gives insight in and easy access to all available statistical data.	2. A metadata system will never be sufficient to give staff the insights they need to use a new source. They should not have easy access to microdata without prior discussion with experienced persons. Subject-matter competence is very important for understanding administrative data. A metadata system cannot replace subject-matter competence.
3. A central statistical data store for managing all available data of interest, enabling the NSO to (re) use this data to create new data/ new outputs, ..., regardless of the data's source	3. *'regardless of the data's source'*. We stress the importance of subject-matter competence. The staff must understand the administrative sources they are using. For example, the administrative unit behind a business identity number can differ between sources as we have shown earlier. This means that knowledge regarding the source is very important for register-based statistics. The statistical surveys are the building blocks of the production system. We fear that data warehousing reduces the role of the surveys. This can have bad consequences on the competence and commitment of the staff.
4. ...the need for a complete new way of organising the statistical production process, ... Not only systems need change, specifically people must change. They have to learn ... new ways of working	4. We also recommend a new way of working. Instead of only working with their own surveys, the staff should also be aware of the possibilities of the register system and be able to combine the methods they use today with work at the systems level. So instead of a 'complete new way of organising the statistical production process', we build on the present way of working with register surveys and sample surveys and developing better methods where the register system is used efficiently.
5. To create fully integrated data sets for enterprise and trade statistics at micro level: a data warehouse approach to statistics	5. As administrative units differ between different administrative sources, this integration is very difficult – errors will be generated if this is not taken care of properly. The calendar year version of the Business Register with carefully created economic variables is what should be created first. No new data warehouse technique is needed for this.
6. The design and implementation of a Statistical Data Warehouse has a huge impact on an NSO. It means developing new IT systems, using new tools etc. asking for high financial investment. It needs a complete redesign of the statistical production process, moving from single operations to integrated generic statistical production.	6. The huge costs for implementing a statistical warehouse must correspond to huge improvements regarding estimators, statistical quality and efficiency. How will a statistical warehouse based on a new IT technique solve the methodological problems that are discussed in our book? The burden of proof lies on those who advocate this new IT technique. In our experience, current problems regarding quality and efficiency are not due to IT problems. Improving quality and efficiency requires new statistical methods and a new systems-oriented paradigm. Both managers and staff should share this new paradigm.

If new data warehousing could facilitate the implementation of the methods we discuss in our book, then data warehousing would be something good.

We are doubtful – a difficult and costly transition from the present IT technique into a new kind of data warehousing will probably disturb the transition from a traditional statistical system based mainly on question-naires and interviewers into the new kind of register-based system that is described here.

The vision of the Commission[3]

The present situation is described as a 'stove pipe model', where each survey produces statistics independently of other surveys. The whole production process from survey design via data collection and processing to dissemina-tion takes place independently for each survey. This stove pipe model has a number of disadvantages: heavy burden on respondents, difficulties with covering multiple dimensions, inefficiency and high costs, redundancies and duplication of work. Users increasingly need integrated and consistent data, and the stove pipe model is not suitable for these demands.

The Commission suggests a transition into a new production system based on modern ICT tools, and there is a clear drive to maximise their use and to gear statistical methods towards them. Statistics for specific domains are no longer produced independently from each other; instead, they are produced as integrated parts of comprehensive production systems for clusters of statistics, the so-called data warehouse approach.

Efficiency gains can be obtained by the reuse of administrative data for sta-tistical purposes. This implies a change in the professional paradigm from 'data collectors' to 're-users of data'. Standardisation and integration of for-merly separated production processes will demand great efforts.

Furthermore, the quality assessment of statistics will become much more complex. Traditional quality measures (e.g. sampling error) will become less relevant as data collection makes less use of sampling techniques. Therefore, a new quality assessment methodology must be developed.

Staff qualifications will need to match the new requirements. New staff with different qualifications must be recruited and current staff will be required to participate in advanced learning.

A comparison between the Nordic system and the Commission's vision

There are many similarities between the reality of the system in the Nordic countries and the Commission's vision. Efficient use of administrative sources is the main issue. Multiple dimensions are covered by creating integrated reg-isters where data from different sectors are combined. The register system can be used to analyse and improve consistency. Sampling errors are less relevant, and the register system can be used to describe non-sampling errors.

However, there are some differences. We do not believe that improvements can be achieved with a massive and costly transition into a completely new produc-tion system based on new IT tools. Instead, new statistical methods for designing systems of surveys should be developed. The subject-matter competence of the present staff is important and should be strengthened and combined with an understanding of the possibilities offered by the register system.

---

[3]This section is based on abstracts from the report by the Commission of the European Communities (2009).

12.5.3   Missing values – a system-based approach

The first cell on the last row in the table in Section 12.5.2 is repeated here. This section explains what we consider is an example of 'integrated generic statistical production' regarding how missing values should be handled.

6. The design and implementation of a Statistical Data Warehouse has a huge impact on an NSO. It means developing new IT systems, using new tools etc. asking for high financial investment. It needs a complete redesign of the statistical production process, moving from single operations to **integrated generic statistical production**	We rewrite Goossens' text in our register-statistical terms: The design and implementation of a register-based statistical system has a huge impact on an NSO. It means developing new survey methods, moving from a *one-survey-at-a-time* paradigm into a *system-based* paradigm. The new production system will be more cost-effective and give more opportunities and improved quality.

The expensive new IT systems and IT tools come first in Goossens' paper and in the Commission's vision, whereby statistical methodology must adapt accordingly. We do not understand this; new survey methods must be developed first, and thereafter the development of the requisite IT systems and IT tools.

We illustrate this way of working with an example with missing values in 'a cluster' of statistics consisting of the Quarterly and Yearly Pay Registers, the Business Register and the National Accounts.

Section 6.3 discusses consistency editing; the discussion is illustrated by an example with the Quarterly Pay Register, the Annual Pay Register, the Business Register and the National Accounts Survey.

From Chart 6.3b   After matching the Business Register and the Annual Pay Registers

(1) BIN	(2) Sector	(3) ISIC	(4) BIN	(5) WagesYear	(6) W-imp		Name	Count	Missing
*	*	*	BIN01	25	0	(1)	BIN	365 061	33 543
BIN02	6	52	*	*	*	(2)	Sector	365 061	33 543
BIN03	1	51	BIN03	1 667	0	(3)	ISIC	365 061	33 543
BIN04	7	91	BIN04	796	0	(4)	BIN	365 061	59 650
BIN05	1	28	BIN05	2 000	1	(5)	AggrWages	365 061	59 650
BIN06	1	45	BIN06	92	0	(6)	W-imp Imputed values, wages		
BIN07	1	51	BIN07	4 758	0				
BIN08	1	60	BIN08	39	0				
BIN09	1	28	BIN09	452	0				
BIN10	1	74	*	*	*				
BIN11	1	27	BIN11	289	0				
...									

A large amount of *non-match* is found when the 331 518 observations with active employers in the Business Register are matched with the 305 411 observations in the administrative Annual Pay Register. There are 33 543 observations in the administrative register that are missing in the Business Register.

The non-match will generate missing values in the variables *institutional sector* and *industry*. In addition, a yearly wage sum of about SEK 6 616 million is not linked to any industry in the tables that the Annual Pay Register sends to the National Accounts representing about 0.7% of the total wage sum.

At the Annual Pay Register, this problem is regarded as so small that it can be disregarded. This means that the staff at the National Accounts must solve the problem, and they probably 'smear' the 6 616 million on other estimates so that the problem seems to disappear.

This is the traditional approach to missing values. A system-based approach (or 'integrated generic statistical production method') can be described as follows:

- Adjustment for missing values in register surveys should be *coordinated and consistent*. In the example with the Quarterly and Annual Pay Registers these two surveys should not be adjusted independently.

- *Clear responsibilities*. In sample surveys, each survey is responsible for making adjustments for missing values in their own survey, but the same strategy will not work in register surveys. The non-match in the wage sum example is caused by undercoverage in the Business Register. As one of the base registers, the Business Register is responsible for coverage errors and for the quality of the standardised variables in the Business Register. In the example with wage sums, the Business Register should reduce undercoverage (by using sources with activity variables; see Chart 10.11). The Business Register should also adjust for missing values in the *sector* and *industry* variables; then all surveys should use these adjustments.

- Missing values in qualitative variables should be handled *with random imputations*.[4] Within each sector, random imputations can be used with a nearest neighbour method based on the wage sum variable. More variables as, e.g. turnover, can be imported in the Business Register and be used in a more advanced imputation model.

The problems with missing values in a system with four important surveys can be solved with the method outlined here. The method should be used every quarter, and more economic registers and surveys can be added to the system. The method requires no expensive IT systems and can be handled by the present staff at the NSO.

---

[4] More on random imputation can be found in Chapter 12 in the second edition of our book, Wallgren and Wallgren (2014).

12.6   Final remarks

We have selected the following three charts, Charts 2.10, 2.11 and 2.14 as the most important charts in the book. Why?

12.6.1   The Statistical Population Register

Charts 2.10 and 2.11 highlight the important issue that many new register countries must deal with – the work with improving the quality of the administrative registers that will be used for social statistics in the country.

Section 12.1 notes that the first register-based census 1981 was met with distrust by many statisticians in Europe. In many member states, the statisticians had compared their population registers with census data and found errors in register data of about 10–20% of the population in small areas. This was the reason for their distrust. We have also met similar attitudes: 'Our Registro Civil is worthless, register-based population statistics are not possible!'

If the administrative population register is bad today, then the situation in the country can probably be described by Chart 2.10 – many bad population registers and uncoordinated work by the ministries and authorities in the country is the rule.

How to improve the quality? Cooperation is the method. *One* authority is given the role of being responsible for *the* administrative Population Register in the country. All other authorities do not produce their own registers exclusively – they all start with the central Population Register. As this register now is being used much more frequently, the quality will gradually improve as the users report the errors they find.

Chart 2.10   Uncoordinated work with registers

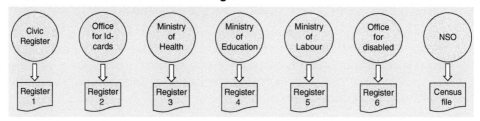

Chart 2.11   Cooperation regarding the central population register

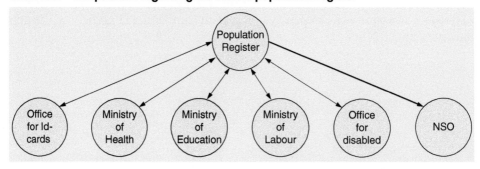

Chart 2.14   Sweden's system 2011

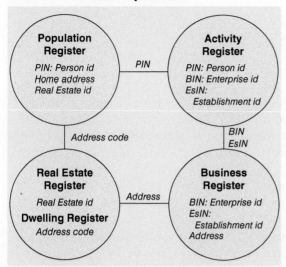

12.6.2   The system of base registers

With the system of base registers, persons and jobs can be georeferenced. This is one very important feature of the system of base registers.

Another important feature is that all registers can be combined with all other registers and all sample surveys can be combined with all registers. This makes new kinds of quality assessment possible and all NSOs should explore these new possibilities.

References

Argüeso, A. and Vega, J. (2013) A population census based on registers and a '10% survey'. Paper presented to the 59th ISI World Statistics Congress, Hong Kong, Session STS063.

Biemer, P. (2010) Total survey error – design, implementation and evaluation. *Public Opinion Quarterly*, **74**(5), 817–848.

Bycroft, C. (2015) Census transformation in New Zealand: Using administrative data without a population register. *Statistical Journal of the IAOS*, **31**(3), 401–411.

Cain, D., Figueroa, V. and Herrera, S. (2019) Hacia un sistema de estadísticas basadas en registros administrativos: una propuesta metodológica para evaluar registros administrativos. *Revista de Estadística y Metodologías, Número 5*.

Chieppa, A., Gallo, G., Tomeo, V., Borrelli, F. and Di Domenico, S. (2018) Knowledge discovery for inferring the usually resident population from administrative registers. In Mathematical Population Studies. (Online) Journal homepage:http://www.tandfonline.com/loi/gmps20.

Cochran, W. G. (1963) *Sampling Techniques*, 2nd edn. New York: Wiley.

Commission of the European Communities (2009) *Communication from the Commission to the European Parliament and the Council on the production method of EU statistics: a vision for the next decade*. Brussels 10. 8.2009 COM(2009) 404 final.

Copete, A. F., Sanchez, C. and Recaño, J. (2016) Proposed methodology for estimating international migration. Conference of European Statisticians, Work session on migration statistics, Geneva May 2016.

Cruz, J. and Cerrón, M. (2015) The build of a statistical register based on administrative data for the production of international migration statistics in the Andean Community countries. Proceedings of the ISI World Statistics Congress 2015.

Daas, P., Ossen, S., Tennekes, M., Zhang, L.-C., Hendriks, C., Foldal Haugen, K., Cerroni, F., Di Bella, G., Laitila, T., Wallgren, A. and Wallgren B. (2011) *Report on methods preferred for the quality indicators of administrative data sources*. Second deliverable of work package 4 of the BLUE Enterprise and Trade Statistics project.

Daas, P., Ossen, S. and Tennekes, M. (2012) Quality report card for administrative data sources including guidelines and prototype of an automated version. Third deliverable of work package 4 of the BLUE Enterprise and Trade Statistics project.

Dargent, E., Lotta, G., Mejía, J. A., Moncada, G. (2018) *Who wants to know? The Political Economy of Statistical Capacity in Latin America*. Inter-American Development Bank, Washington D.C.

Dasylva, A., Goussanou, A., Ajavon, D. and Abousaleh, H. (2019) *Revisiting the probabilistic method of record linkage*. Preprint: Statistics Canada and University of Victoria, Researchgate.

De Waal, T. (2009) Statistical data editing. In D. Pfefferman and C.R. Rao (eds), *Sample Surveys: Design, Methods and Applications, Handbook of Statistics Volume 29A*, pp. 187–214. Amsterdam: Elsevier.

De Waal, T., and Quere, R. (2003), A Fast and Simple Algorithm for Automatic Editing of Mixed Data. *Journal of Official Statistics*, **19**, 383–402.

Deville, J. and Särndal, C-E. (1992) Calibration estimators in survey sampling. *Journal of the American Statistical Association*, **87**, 376–382.

Dias, C., Wallgren, A., Wallgren, B., and Coelho, P. (2016) Census Model Transition: Contributions to its Implementation in Portugal. *Journal of Official Statistics*, **32**(1), 93–112.

Eltinge, J., Biemer, P. and Holmberg, A. (2013) A potential framework for integration of architecture and methodology to improve statistical production systems. *Journal of Official Statistics*, **29**(1), 125–145.

Eurostat (1997) *Proceedings of the Seminar on the Use of Administrative Sources for Statistical Purposes, 15–16 January 1997*. Luxembourg: Office for Official Publications of the European Communities.

FAO (2010) Global Strategy to improve Agricultural and Rural Statistics. Report number 56719-GLB. The World Bank 2010.

Ferraz, C. (2015) *Linking Area and List Frames in Agricultural Surveys*. Paper written for FAO.

Fellegi, I. P. and Holt, D. (1976). A systematic approach to automatic edit and imputation. *Journal of the American Statistical Association*, **71**, 17–35.

Fellegi, I. P. and Sunter, A. B. (1969) A theory for record linkage. *Journal of the American Statistical Association*, **64**, 1183–1210.

Goossens, H. (2013) Building the statistical warehouse to improve statistics. Paper presented to the 59th ISI World Statistics Congress, Hong Kong, Session IPS023.

Granquist, L. and Kovar, J. (1997) Editing of survey data: how much is enough? In L.E. Lyberg, P. Biemer, M. Collins, E.D. de Leeuw, C. Dippo, N. Schwarz and D. Trewin (eds), *Survey Measurement and Process Quality*, 416–435. New York: Wiley.

Groves, R. (1987) Research on Survey Data Quality. *Public Opinion Quarterly*. 1987; **51**(4): S156–S172.

Groves, R., Fowler, F., Couper, M., Lepkowski, J., Singer, E. and Tourangeau, R. (2004) *Survey Methodology*. New York: Wiley.

Groves, R. and Lyberg, L. (2010) Total survey error – past, present and future. *Public Opinion Quarterly*, **74**(5), 849–879.

Haldorson, M. (2019) *High demand for local area level statistics – How do National Statistical Institutes respond?* Regional Statistics, Vol 9. *No.* **1**. 2019: 168–186; DOI: 10.15196/RS090106

Hand, D. (2018) *Statistical challenges of administrative and transaction data. J. R. Statist. Soc. A.* **181**, Part 3, 555–605.

Herzog, T., Scheuren, F. and Winkler, W. (2007) *Data Quality and Record Linkage Techniques*. New York: Springer.

Hoogland, J., van der Loo, M., Pannekoek, J. and Scholtus, S. (2011) *Data editing – Detection and correction of errors, Statistical Methods (201110)*. The Hague: Statistics Netherlands.

IAOS (2020) Statistical Journal of the IAOS https://officialstatistics.com/news-blog/population-censuses-are-statistical-dinosaurs-able-adapt

Jorner, U. (2008) *Summa summarum – SCB: sförsta 150 år (In Swedish)*. Statistics Sweden, Örebro.

Kardaun, J.W.P.F. and Loeve, J.A. (2005) Longitudinal analysis in statistical offices. Statistics Netherlands Discussion Paper 05010.

Lange, A. (1993) *How do we get register statistics to be recognized in Europe?* Statistics without Frontiers, The 19th Conference of Nordic Statisticians in Reykjavík 1992 (in Danish).

Laitila, T., Wallgren, A. and Wallgren B. (2012) *Quality Assessment of Administrative Data – Data Source Quality.* Part two of third deliverable of work package 4 of the BLUE Enterprise and Trade Statistics project.

Lenk, M. (2009) *Methods of Register-based Census in Austria.* Paper presented at the Seminar on Innovations in Official Statistics, United Nations, New York, February. http://unstats.un.org/unsd/statcom/statcom_09/seminars/innovation/innovations_seminar.htm

Lohr, S. (2009) Introduction to Part 1: Sampling and survey design. In D. Pfefferman and C.R. Rao (eds), *Sample Surveys: Design, Methods and Applications, Handbook of Statistics Volume 29A*, pp. 3–8. Amsterdam: Elsevier.

MacDonald, Alphonse L. *Of science and statistics: The scientific basis of the census.* Statistical Journal of the IAOS 36 17–34. DOI 10.3233/SJI-190596, IOS Press 2020.

Nordbotten, S. (1967) *Purposes, Problems and Ideas Related to Statistical File Systems.* Proceedings from the 36th Session of the International Statistical Institute. Invited paper, Sydney (1967). Available for free downloading from www.nordbotten.com

Pannekoek, J. and de Waal, T. (2005) *Automatic edit and imputation for business surveys: The Dutch contribution to the EUREDIT project. Journal of Official Statistics*, **21**(2), 257–286.

Särndal, C.-E. and Lundström, S. (2005) *Estimation in Surveys with Nonresponse.* Chichester: Wiley.

Selander, R. (2008) *Comparisons between the Business Register and Farm Register in Sweden.* Paper presented at the First International Workshop on Technology and Policy for Accessing Spectrum, TAPAS 2006.

Selander, R., Svensson, J., Wallgren, A. and Wallgren, B. (1998) *Administrative Registers in an Efficient Statistical System – New Possibilities for Agricultural Statistics? How Should We Use IACS Data?* Statistics Sweden and Eurostat.

Statistics Denmark (1994) *Personstatistik i Danmark – Et registerbaseret statistiksystem.* Danmarks statistik, Köpenhamn.

Statistics Denmark (1995) *Statistics on Persons in Denmark – A register-based statistical system.* https://op.europa.eu/da/publication-detail/-/publication/69eab1e5-885d-411c-9c2d-8bd295ced243#

Statistics Sweden (2007a) *Register-based Activity Statistics (in Swedish). Background Facts – Labour and Education Statistics* 2007:2.

Statistics Sweden (2007b) *Register-based economic statistics based on a standardised register population – A calendar year version of the business register with consistent microdata designed for the yearly National Accounts (in Swedish). Background Facts – Economic Statistics* 2007: 6

Statistics Sweden (2013) The Sampling and Estimation Procedure in the Swedish Labour Force Surveys 2005– (in Swedish). *Background facts Labour and Education Statistics* 2013:6.

Statistics Sweden (2017) Analysis on nonresponse bias for the Swedish Labour Force Surveys (LFS) – (in Swedish). *Background facts Labour and Education Statistics* 2017:1.

UN/ECE (2007) *Register-based statistics in the Nordic countries – Review of best practices with focus on population and social statistics.* United Nations, Geneva.

Wallgren, A. and Wallgren, B. (2010) Using administrative registers for agricultural statistics. In R. Benedetti, M. Bee, G. Espa and F. Piersimoni (eds), *Agricultural Survey Methods*, pp.27–44. Chichester: Wiley.

Wallgren A., and Wallgren B. (2014): *Register-based Statistics – Statistical Methods for Administrative Data*, 2nd edn. Chichester: Wiley.

Wallgren A., and Wallgren B. (2020): *Comments on the scientific basis of the census.* Statistical Journal of the IAOS **36** (2020) 1295–1297 1295 DOI 10.3233/SJI-200736.

Winkler, W. (1995) Matching and record linkage. In B. Cox, D. Binder, N. Chinnappa, A. Christianson, M. Colledge and P. Kott (eds), *Business Survey Methods.*, 355–384. New York: Wiley.

Winkler, W. (2006) Overview of Record Linkage and Current Research Directions. Research Report Series 2006/2, US Bureau of the Census.

Winkler, W. (2008) Record linkage. In D. Pfefferman and C.R. Rao (eds), *Sample Surveys: Design, Methods and Applications, Handbook of Statistics Volume 29A*, pp. 351–380. Amsterdam: Elsevier.

Index

Accuracy, quality indicators, 49,
    89–94, 102–104, 128
Activities or Activity data, 37, 45, 71,
    99, 132, 169, 197–198, 232, 234
Adjoined variable, 66, 70f, 241
Administrative object set, 7, 32, 57,
    91, 117–118, 177
Administrative object types or units,
    11, 56, 91, 189, 192–193, 236
Administrative register, 7
Aggregated variable, 70f
Aggregation error, 202–209, 213–217
Anonymous identity numbers, 27,
    33, 36–37, 152, 188
Area frames, area sampling, 3–5, 25,
    82, 96
Automatic editing, 34, 109–110
Auxiliary variable, 97, 156, 159–165

Base register, 3, 46–48, 56–57, 93, 98,
    100, 134, 254
Basic editing, 31, 109
Big data, 5–6
Blocking, 76
Business intelligence, 56

Calendar year register, 138, 140,
    179f, 184, 197, 203–204, 219–220
Calibration, 152, 155f, 161f, 163f
Census, 3, 227f
Centralised system, 5, 47, 188, 218
Changing register, 134
Classification error, 150, 243f
Coding, 11, 104, 190

Coherence, 98
Cohort, 236–237
Combination object, 194, 201, 207f,
    215f, 222f
Communication variable, 28,
    135, 198
Confidentiality, 15, 26f, 134, 188
Consistency, 19, 47, 98f
Consistency editing, 31, 107–108,
    110f, 249
Coordinated system, 19, 98
Coverage problems, 6, 117, 126, 131,
    149f, 182
Current stock register, 137

Data matrix, 8
Data warehouse, 250–252
Demographic event, 38–39, 44, 131,
    133–134, 136–139, 144
Derived object or unit, 13, 119–120,
    194, 247
Derived variable, 13, 62, 70–73, 122,
    241–244
Decentralised system, 5, 47, 188, 218
Deterministic record linkage, 66, 68,
    82–83, 95
Duplicates, 33–34, 66, 76, 92, 141
Dwelling register, 41, 46–47, 100
Dwelling household, 13, 43, 46, 100,
    146–147

Electricity meter register, 44, 46, 48,
    98, 100, 147
Events register, 139

Register-based Statistics: Registers and the National Statistical System, Third Edition. Anders Wallgren and Britt Wallgren.
© 2022 John Wiley & Sons Ltd. Published 2022 by John Wiley & Sons Ltd.

Flow variable, 140, 171, 179–180
Foreign key, 92, 135
Frame population, 3, 129–131,
　　137, 173

Geocoded statistical units, 232, 234
Georeferencing, 100, 132, 150–152

Historical register, 139

Identifying variable, 28, 31, 65, 68,
　　92, 135
Identity database, 30–37
Identity number principle, 20
Imputation, 104, 109–110, 127,
　　252–253
Input data quality, 86, 90–94, 102,
　　105, 144, 151, 163
Input database, 27–28, 31, 33
Integration error, 203, 247
Inventory of sources, 87, 131,
　　146, 195

Links, 47, 66, 69–70, 78–79
Longitudinal quality, 191, 236, 246
Longitudinal register, 66–67, 139,
　　232, 234, 236–237, 246

Matching error, 69, 73, 75, 79–80,
　　111
Matching key, 65–66, 68–69, 73, 77,
　　79, 91–92, 233
Measurement errors, 16, 51,
　　57, 60f
Metadata, 11–12, 30–32, 55, 87,
　　91, 240
Missing values, 57–58, 92, 113,
　　126–127, 157–159, 178–179,
　　252–253
Model errors, 62, 241–244
Multi-valued variables, 201–218

National statistical system, 2–3, 19,
　　22–26, 40, 44, 48, 100, 142, 188,
　　218, 230–232
Nonresponse bias, 156, 161–163
Non-sampling errors, 17, 128, 251

Object set, 7, 11, 31, 39, 91, 117,
　　172–173, 177–178, 193–194
Object type, 11, 22, 47, 56, 65, 70f,
　　133, 172, 188f
Observation, 8–9
One survey at a time paradigm, 86,
　　114, 238
One-way traffic, 23, 39
Output data quality, 85, 94, 96, 152
Overcoverage, see coverage problems

Paradigm, 14, 16, 86, 109, 184, 227,
　　229, 248, 250–252
Parsing, 75
Primary key, 92, 102, 135
Privacy, 15, 20, 23, 26, 29, 37–38
Probabilistic record linkage, 16, 34,
　　67, 73f, 80, 82, 142, 233
Production database, 31, 35–37
Production process quality, 85–86,
　　90, 93–94, 105, 152
Production system, 2, 5, 21, 25, 89,
　　105, 125–126, 129, 248f

Quality assessment, 14, 17, 19, 30,
　　85f, 107, 117, 125–126, 140, 143,
　　145, 151, 233, 238
Quality indicators, 32, 90–105, 108,
　　111, 144
Quality of estimates, 94, 96, 126

Record, 8–9
Record linkage, 65f, 82–83, 141
Reference variable, 13, 65, 91, 135
Register
　　Administrative register, 7
　　Base register, 3, 46–48, 56–57, 93,
　　　　98, 100, 134, 254
　　Calendar year register, 138, 140,
　　　　179f, 184, 197, 203–204,
　　　　219–220
　　Changing register, 134
　　Current stock register, 137
　　Electricity meter register, 44, 46,
　　　　48, 98, 100, 147
　　Events register, 139

Historical register, 139
Longitudinal register, 66–67, 139, 232, 234, 236–237, 246
Register at a specific point in time, 138, 179
Statistical register, 1, 57
Unchanging register, 133
Register referring to a specific point in time, 138, 179
Register maintenance survey, 102–103, 189–190, 197
Register population, 130–131, 171, 173
Register survey, 1, 10–13
Register system, 2–7, 12, 17, 19, 42–45, 56, 98–100
Registration data, 184, 189, 196–197
Relational object, 234
Relevance, relevance error, 91, 174f

Sampling error, 15, 17, 237–238, 247
Signs of life, 131–132, 143, 145, 152, 167–169
Single-valued variable, 201, 205–206, 209
Small area estimation, 18, 41, 229f, 146, 229, 231
Spanning variable, 88, 135
Standardisation, 74, 230
Standardised population, 66, 129–130, 134, 183, 187, 224
Standardised variable, 129, 135–136, 187, 198, 238, 253
Standardising variable names, 33
Statistical data, 36, 51, 57, 60–61
Statistical register, 1, 5, 7
Stock variable, 140, 171, 179–180
Survey design, 3, 48–49, 90, 125–127, 227, 237
Survey system design, 3, 17, 48, 66, 89, 105
System
  National statistical system, 2–3, 19, 22–26, 40, 44, 48, 100, 142, 188, 218, 230–232
  Production system, 2, 5, 21, 25, 89, 105, 125–126, 129, 248f

Register system, 2–7, 12, 17, 19, 42–45, 56, 98–100
System approach, 7, 16–17, 66–67, 232
System of base registers, 42–48, 232–233, 249
System of surveys, 2, 105, 128
System principle, 19, 98

Target population, 148, 171, 173–175, 177–179, 218
Technical variable, 135
Throughput database, 27–28, 31, 33, 35, 37, 108
Time reference, 11, 92, 129–139
Time series quality, 215, 246
Timeliness, 50
Total survey error, 128
Transformation principle, 18–19, 114, 176, 182

Unchanging register, 133
Undercoverage, see coverage problems
Unit error or problem, 119–120, 194, 224
Updating registers, 31, 136f, 198

Variable
  Adjoined variable, 66, 70f, 241
  Aggregated variable, 70f
  Auxiliary variable, 97, 156, 159–165
  Communication variable, 28, 135, 198
  Derived variable, 13, 62, 70–73, 122, 241–244
  Flow variable, 140, 171, 179–180
  Foreign key, 92, 135
  Identifying variable, 28, 31, 65, 68, 92
  Matching key, 65–66, 68–69, 73, 77, 79, 91–92, 233
  Multi-valued variable, 201–218
  Primary key, 92, 102, 135
  Reference variable, 13, 65, 91, 135
  Single-valued variable, 201, 205–206

Spanning variable, 88, 135
Standardised variable, 129,
    135–136, 187, 198, 238, 253
Stock variable, 140, 171,
    179–180
Technical variable, 135
Time reference, 11, 92, 129–139

Weight-generating variable,
    179–180, 209

View, 9

Weights, calibrated, 96, 152, 155–169
Weight-generating variable, 179–180,
    209